C0-AUQ-106

Elementary
Wave Optics

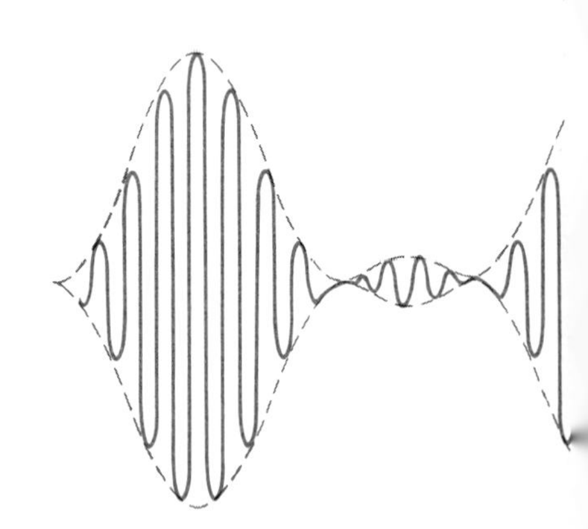

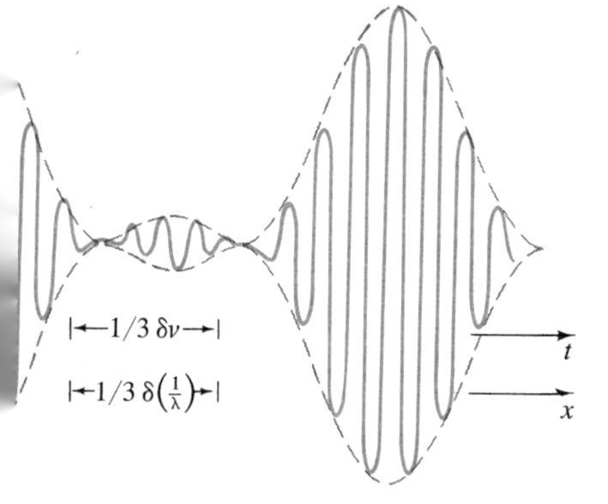

Elementary Wave Optics

ROBERT H. WEBB

Academic Press *New York and London*

ACADEMIC PRESS, INC.
111 Fifth Avenue, New York, New York 10003

United Kingdom Edition published by
ACADEMIC PRESS, INC. (LONDON) LTD.
Berkeley Square House, London W.1

LIBRARY OF CONGRESS CATALOG CARD NUMBER: 72–79896

PRINTED IN THE UNITED STATES OF AMERICA

I'm cariad i.
Yn gyntaf rhaid i fardd ddysgu canu.

Contents

Preface

Wave optics is the study of the various phenomena associated with the wave properties of light—and by extension, with any waves. A knowledge of this subject is necessary for an understanding of optical devices, but more important, a study of wave optics provides a way to introduce many of the phenomena of modern quantum physics. The feature which distinguishes modern physics from classical physics is the occurrence of wave (interference) effects in the interactions of all kinds of matter and energy. This book is designed to explain these phenomena in terms of electromagnetic waves (including visible light) and of other wave systems closer to common experience. Extensively described are superposition, scattering, and the concept of coherence. Coherent optics (so-called "modern optics") has newly become accessible to experimentation, and the approach to modern physics which is emphasized here permits the treatment of this subject in some detail.

In order to maintain the thread of the argument, the main part of the text uses as little formal mathematics as possible. For more advanced students, the appendices will supply the missing derivations, after the subject has been introduced in the text.

Problems are an integral part of this text. Detailed solutions are given to half of them, so that the student can use solved problems as a learning device. Self-restraint in looking up the solution is important, since the value of a problem lies in trying to solve it and in understanding its difficulties. Subsequent use of the solution then makes pedagogic sense. The unsolved problems are roughly paired with the solved ones in most cases. Short exercises of a more conventional sort are also included.

I am grateful for the many contributions of problems, criticisms, and insights from my colleagues in the courses which generated this book. Professors R. F. Walker, J. T. Tessman, and A. E. Everett, in particular, will recognize their efforts—I hope faithfully included. Finally, and in dedication, this is for my most important luminary—Charme.

ROBERT H. WEBB

Lexington, Massachusetts

Elementary
Wave Optics

1 Geometrical optics: summary

Geometrical optics deals with light (and more generally with waves) in situations where it is possible to ignore the wave character of the phenomenon. This usually means that the wavelengths involved are very much smaller than the dimensions of anything with which the waves interact. Later we will see that geometrical optics is a limiting case of wave optics. The usual value in limiting cases is their simplicity, and geometrical optics shares this asset, with reservations.

All of geometrical optics may be deduced from three simple empirical rules:

1. Light travels in straight lines in homogeneous media.
2. The angle at which light is reflected from a surface is equal to the angle at which it is incident.
3. When light passes from one medium to another, its path is described by the equation $n_1 \sin \theta_1 = n_2 \sin \theta_2$.

Figure 1.1 summarizes these rules and defines the various angles. A "ray" is a line along the path the light follows. We think of this as a very narrow beam of light.

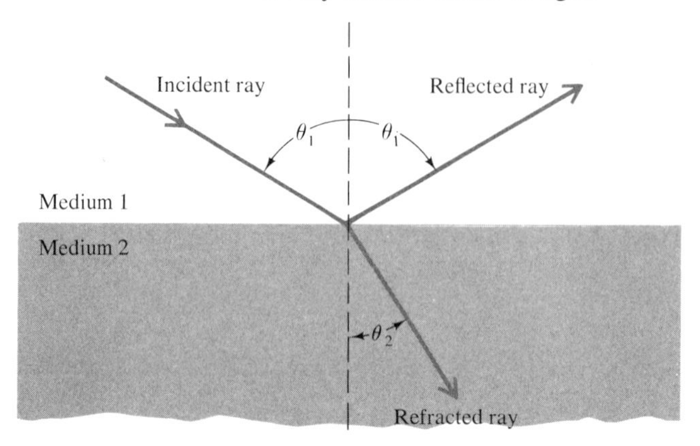

Figure 1.1: Reflection and refraction.

We may regard rule 3, which describes *refraction*, as defining the relative *index* of refraction: n_2/n_1.

If we follow custom and define the index of vacuum to be $n_{\text{vac}} = 1$, then n_1 and n_2 are the indices of each medium relative to vacuum, and are so listed in handbooks. Later we will see that this ability to characterize each medium by a single number is extremely important. Among other consequences, it will lead us to regard n_k as the ratio of the speed of light in vacuum (c) to (c_k), the speed of light in medium k: $n_k = c/c_k$. This in turn allows us to deduce the three rules from the more general Fermat's principle. The main asset of this principle is the esthetic one of unification. It is not essential to the conclusions of geometrical optics, although occasional simplifications are possible. But the three rules suffice.

A further statement limits the kinds of media usually considered. This is the *reciprocity* principle, which requires that if light can follow a certain path from A to B, then it can follow the same path from B to A. Some media do not support this principle, but they seldom occur in questions pertaining to geometrical optics.

If we apply our rules directly to plane surfaces, they describe the behavior of mirrors and prisms (1.1–1.3)*. Rule 3 also predicts the phenomenon of *total internal reflection*. When light passes from a medium with a larger index to one with a smaller index (as, for

* Numbers in parentheses indicate relevant problems at the end of the chapter.

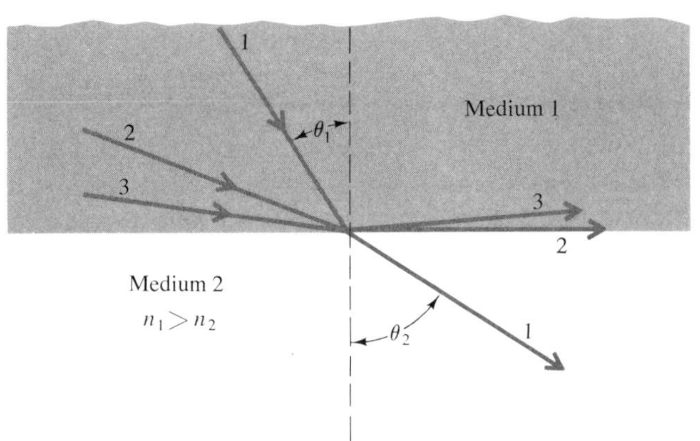

Figure 1.2: Total internal reflection.

instance, from glass to air), the ray is bent toward the surface. That is, $\theta_2 > \theta_1$. Eventually, the ray emerging from the "denser" medium (the one with larger index) lies parallel to the surface. This occurs at the critical angle: $n_1 \sin \theta_c = n_2 \sin(90°) = n_2$. If θ_1 is bigger than θ_c, no light emerges. In this case, all incident light is contained in the reflected ray so that, from inside, the surface appears to be a perfect mirror. Such mirrors are important in various optical instruments. Familiar examples are the right-angle prisms in binoculars and the light pipes which illumine hard-to-reach places.

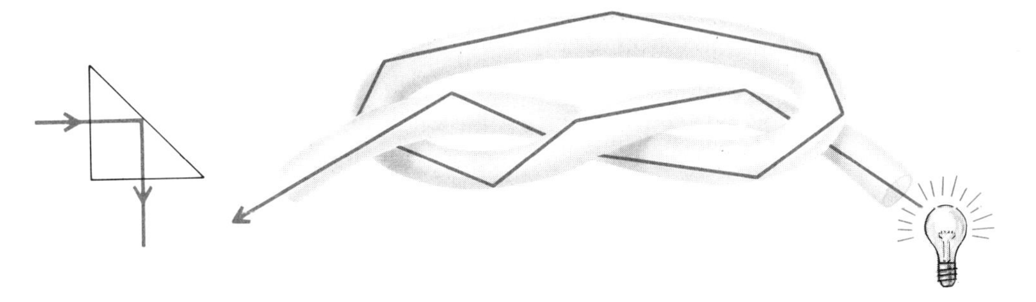

Figure 1.3: Applications of total internal reflection.

Notice that all light incident from the *outside* of a totally reflecting surface will enter the surface, but it will not reach the region for which $\theta_1 > \theta_c$. (1.4)

In applying rule 3, we find a new empirical fact: n is different for different colors of light. Later we will study the source of this *dispersion* in some detail, but in geometrical optics we merely use it, for example, when we separate colors with a prism. Or, we may compensate for it, as in making achromatic lenses of two kinds of glass, in which the dispersion of one compensates for that of the other. (1.5)

Rule 2 for reflection governs the behavior of instruments with curved mirrors, such as the astronomical telescope. Since the manipulation is similar to that of lenses, we can summarize the operations of the two devices together.

A converging lens is a device which brings parallel light to a single point, called the *focus*. "Parallel light" means a beam in which any one ray is parallel to any other in the beam. Such light comes from a source so distant that the divergence of two adjacent rays is imperceptible, or (by the reciprocity principle) it comes from a source at the focal point of a converging lens. The three rules of geometrical optics tell us how to construct a real lens, but we will generally deal with existing ones, deferring consideration of their construction until needed. So, using the definition of a converging lens alone, and assuming an ideal lens, we can find how images are formed.

Without derivation, we present the *thin-lens* equation:

$$\frac{1}{p} + \frac{1}{q} = \frac{1}{f}.$$

This refers to a lens thin enough to have its focal points at equal distances on each side. Practically, the equation works only for rays nearly parallel to the axis of a real lens.

Figure 1.4 defines the parameters. Notice that the way to find the

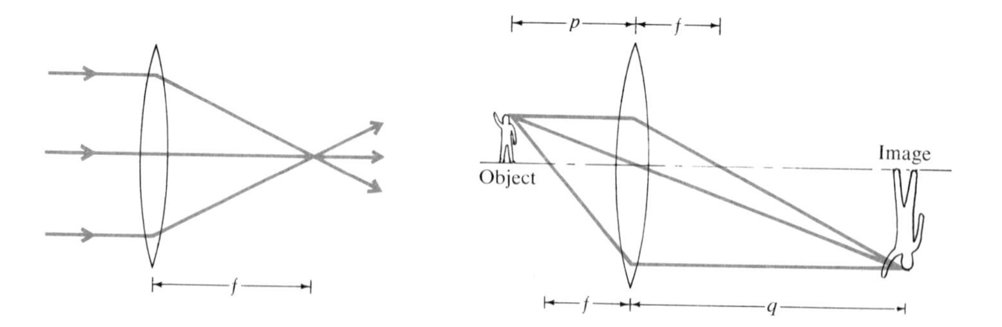

Figure 1.4: Thin-lens parameters.

image graphically is to follow two of the rays diverging from an off-axis point on the object. The one going through the center of the lens is undeflected. (This follows from rule 3, in the approximation of the *thin* lens, which has two close, parallel surfaces at its center.) The ray parallel to the axis must go, by definition, through the far focal point. The ray going through the near focal point becomes parallel to the axis. Two of these three rays are enough. In the situation illustrated, the rays actually come together again, and a *real* image (one which can be cast on a screen) is formed. Subsequently, the rays diverge from a real image just as they do from a real object. The size of the image may be inferred immediately from the geometry. The magnification M is defined by the equation:

$$M = \frac{Y_{\text{image}}}{Y_{\text{object}}} = \frac{-q}{p}.$$

(The negative sign indicates that the image is inverted.) M is called the linear magnification.

If the rays never do meet again, they will appear (by rule 1) to come from a *virtual* image, as shown in Figure 1.5.

Figure 1.5: Virtual images.

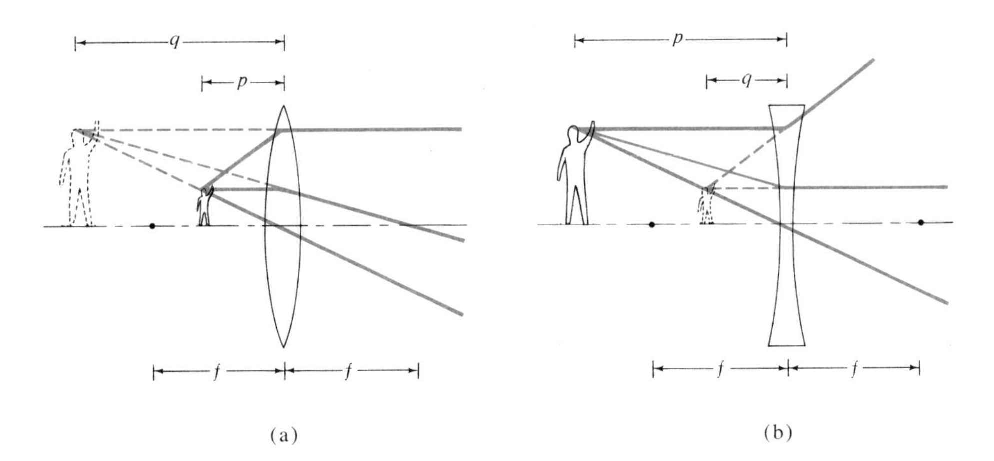

The lens in Figure 1.5(b) is a *diverging* lens, which makes parallel light diverge as if it came from a point. We can describe all thin lenses and all situations by the same equation if we adopt a few conventions to clarify the way in which the equation is used:

1. If the *incident* light comes *from the* OBJECT, we say it is a *real* object, and define the distance from the lens to it as *positive*. Otherwise, it is *virtual*, and its distance is *negative*.

2. If the *emergent* light goes *toward* the IMAGE, we say it is a *real* image, and define the distance from the lens to it as *positive.*

3. The FOCAL LENGTH is *positive* for a *converging* lens or a *concave* mirror and *negative* for a *diverging* lens or a *convex* mirror.

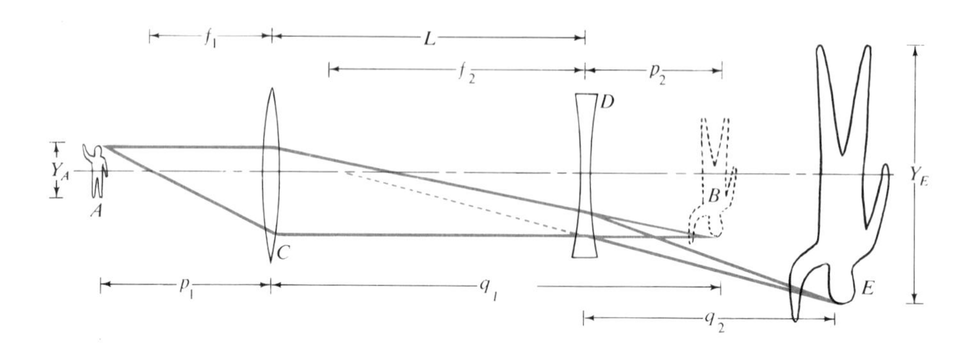

Figure 1.6: Image formation, two lenses.

The following exercise illustrates these conventions: In Figure 1.6, A is a real object for the converging lens C. B is the consequent real image, or would be if the diverging lens D did not intervene. B is also a virtual object for lens D, and E is the consequent real image. Some numerical values might be: $f_1 = +5$ cm; $f_2 = -10$ cm; $p_1 = +7$ cm; $Y_A = +1$ cm; and $L = 12.5$ cm.

First find q_1, the position of B:

$$q_1 = \frac{f_1 p_1}{p_1 - f_1} = \frac{+35}{2} \text{ cm} \qquad \text{(real).}$$

Its size is

$$Y_B = -\left(\frac{q_1}{p_1}\right) Y_A = \frac{-5}{2} \text{ cm} \qquad \text{(inverted).}$$

Since the object distance for the lens D is the difference between q_1 and L we find $p_2 = -5$ cm, with the negative sign indicating that it is on the side of D to which the light goes. Again we apply our equation and find that

$$q_2 = \frac{f_2 p_2}{p_2 - f_2} = \frac{(-10)(-5)}{-5-(-10)} \text{ cm} = 10 \text{ cm.}$$

The object size is

$$Y_E = -\left(\frac{q_2}{p_2}\right) Y_B = -5 \text{ cm} \qquad \text{(inverted).}$$

Mirrors also obey the thin-lens equation, and the conventions apply in the same way. (1.6–1.10)

So far we have said nothing about the actual surfaces from which lenses and mirrors are made, nor have we provided for lenses of finite thickness. In fact, the ideal surfaces are seldom available for lenses and are often unnecessary for mirrors. The ideal surface for a lens is a very complicated curve, derivable from our rules but difficult to fabricate. That for a mirror is the simpler ellipsoid. Since a spherical surface is easiest to make, lenses (and sometimes mirrors) usually have this form.

A spherical mirror has a focal length equal to one-half the radius of curvature. Using this value, it is easy to see that only rays near the axis will be focused sharply.

For a single spherical *refracting* surface, the following equation holds:

$$\frac{n_{\text{inc}}}{p} + \frac{n_{\text{em}}}{q} = \frac{n_{\text{em}} - n_{\text{inc}}}{R},$$

where n_{inc} is the index on the side from which the light is incident, n_{em} is that on the side into which the light emerges, and R (the

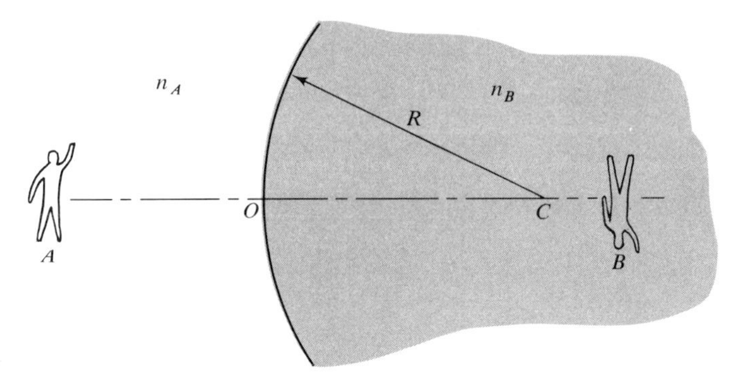

Figure 1.7: Single, spherical-surface parameters.

radius of curvature of the surface) is positive if the center of curvature is on the side toward which the light goes. For example, in Figure 1.7 an object at A yields an image at B according to

$$\frac{n_A}{\overline{OA}} + \frac{n_B}{\overline{OB}} = \frac{n_B - n_A}{\overline{OC}},$$

or an object at B yields an image at A:

$$\frac{n_B}{\overline{OB}} + \frac{n_A}{\overline{OA}} = \frac{n_A - n_B}{-\overline{OC}}.$$

As with lenses, we apply our equation to each surface in turn, ignoring the others. (1.11–1.12)

Keep in mind that this summary is for the simplest of applications of geometrical optics, although its assumptions are sufficient for an understanding of that subject. This explanation should enable you to use reference texts, such as Jenkins and White (see Bibliography), and help you to understand real (for example, thick) lenses, but you must expect to labor diligently to apply such knowledge. Applied geometrical optics as practiced today is one of the most complicated subjects ever derived from three simple rules.

EXERCISES

1. If a 6-ft man can just see from his feet to the top of his head in a mirror, how tall is the mirror?

2. A prism has a small peak angle, α, and is made of glass with index n. Through what angle will it bend a light beam?

3. Find the size and position of the sun's image due to a lens of focal length +50 cm. (The sun subtends an angle of about 0.5 degree.)

4. Find the size and position of the moon's image due to a lens of focal length −50 cm.

5. Use the equation for the image formed by a single spherical surface to find the apparent depth of a penny which is actually under 10 cm of water.

PROBLEMS

These problems are intended to supplement, as well as illustrate, the material covered above. Relevant sections of the text are followed by problem numbers in parentheses. Solutions are given for selected problems at the end of the book.

1.1 Three mirrors are set at right angles to each other to form a "corner reflector." If a light is shone into this system, where does the beam go?

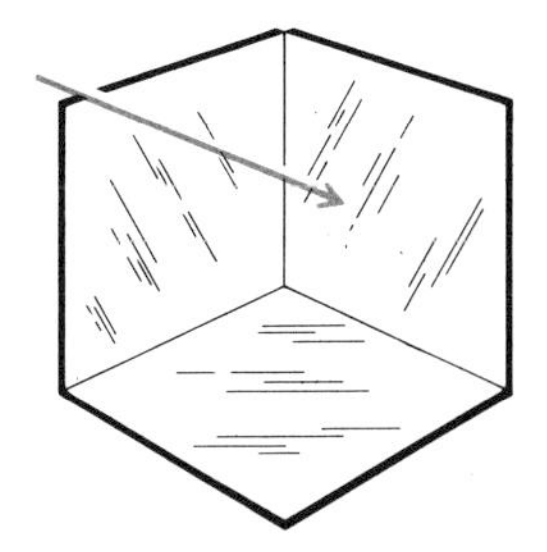

1.2 Where does the light ray in the figure hit the screen? Suppose that at each interface one-half the light intensity goes into the reflected ray and one-half into the refracted (transmitted) ray. Find the second brightest spot on the screen.

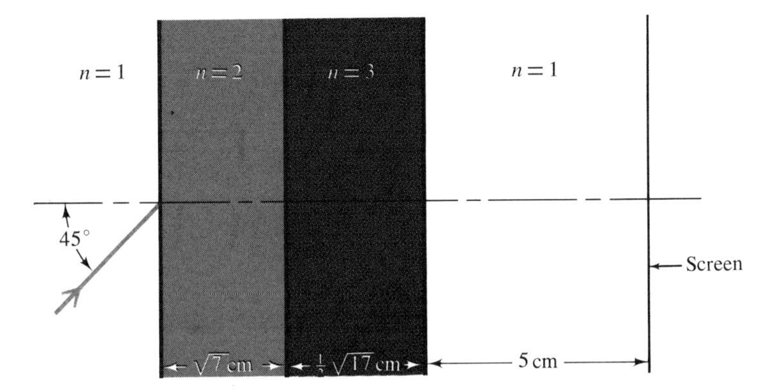

1.3 A fish watcher looks at an aquarium from a point on a line diagonally through it. For a fish on this line of sight, 5 cm from the corner, how many images does the watcher see, and what are their locations. Does the fish appear different in the different images? What does the fish see?

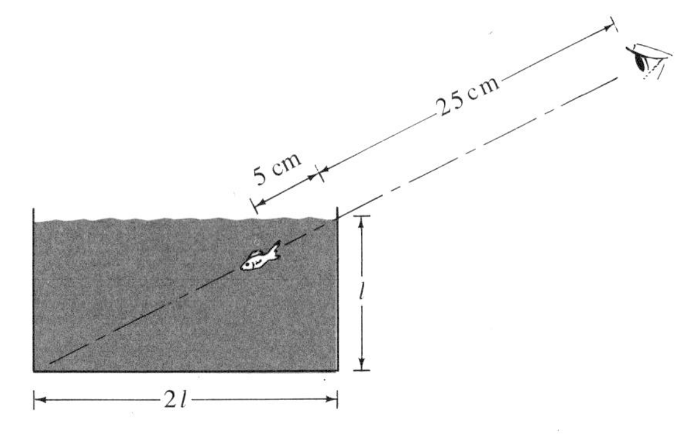

1.4 A skin diver shines his flashlight at the surface on the water so that the beam makes an angle of 60 degrees with the vertical.

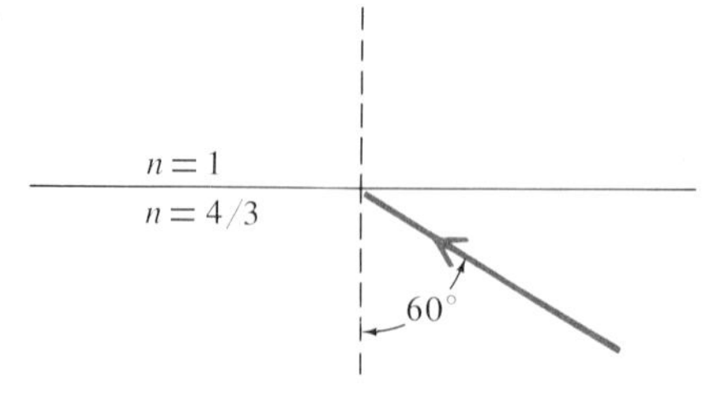

(a) Where does the beam go? Assume that there is no reflected beam if there is a transmitted one.

(b) Oil of index 1.2 is now spread on the water. Where does the beam go?

(c) Many layers of oil are spread on the water, as shown. Sketch the path.

$n = 1.00 \rightarrow$
$1.05 \rightarrow$
$1.10 \rightarrow$
$1.15 \rightarrow$
$1.20 \rightarrow$
$1.25 \rightarrow$
$1.33 \rightarrow$

(d) The air over a blacktop road is hottest near the road surface. The index of air far from the surface is 1.0003. An observer sees the road surface only if he looks down at an angle of 89 degrees or less. What is the index of air at the surface?

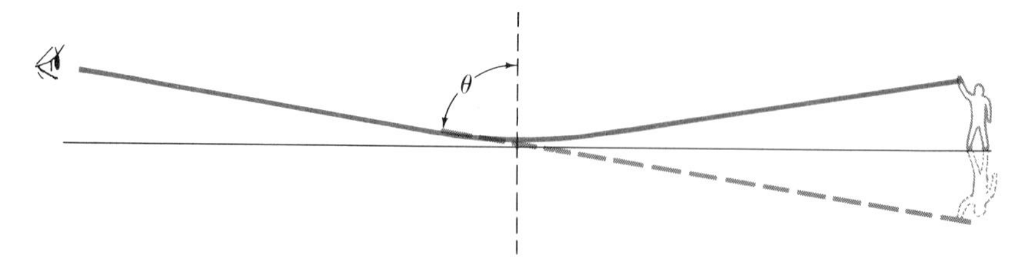

1.5 The index of the glass used to make a prism is 1.55 for red light and 1.65 for blue. Design an arrangement whereby *only* red light emerges from the prism. Take the index of air to be 1.00. *Hint:* Let the beam enter perpendicularly to a surface.

1.6 Find and describe the image in the following lens combinations:

(a) Simple magnifier:

$Y_{obj} = 1$ mm, $p = 5$ cm, $f = 6$ cm.

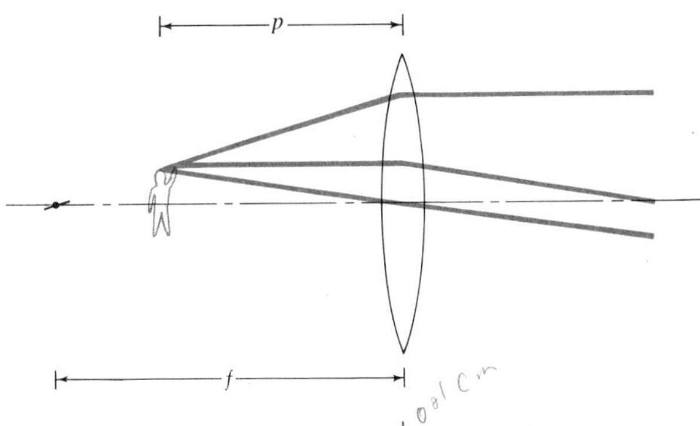

(b) Compound microscope:

$Y_{obj} = 0.01$ mm, $p_1 = 1.1$ cm,

$f_1 = 1$ cm, $f_2 = 10$ cm, $L = 18$ cm.

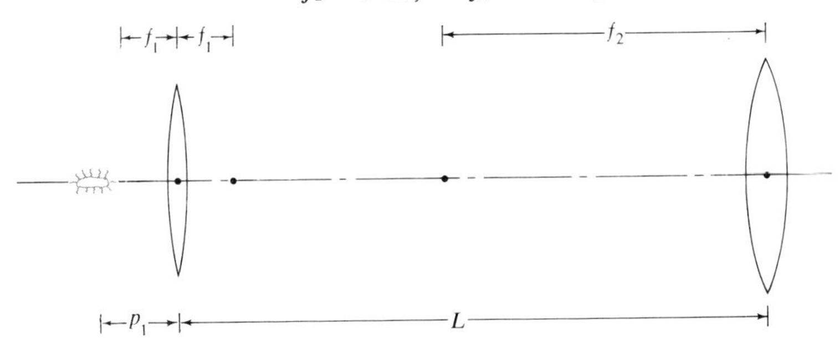

(c) Astronomical telescope:

$Y_{obj} = 2.5$ thousand miles, $p_1 = 0.25$ million miles,

$f_1 = 10x$ miles, $f_2 = x$ miles, $10x <<< 0.25$ million.

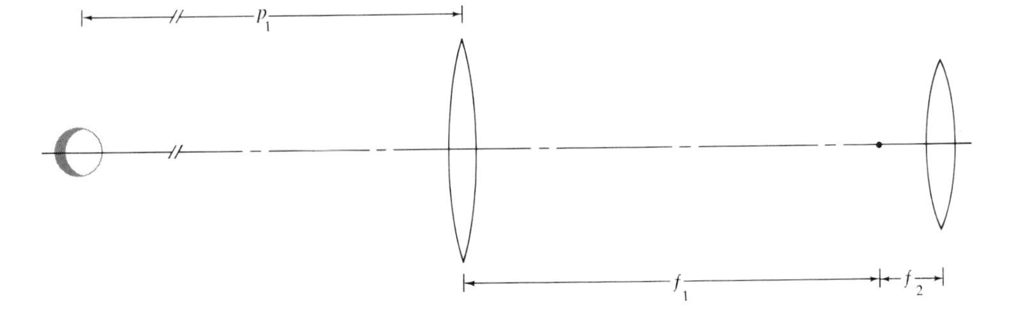

Does the image *appear* bigger or smaller? *Hint:* Keep at least the second term in q_1 before approximating. (See Appendix A for help with approximation.)

(d) Opera glass (Gallilean telescope):

$Y_{obj} = 2\text{m}$,
$p_1 = 50$ m,
$f_1 = 0.10$ m,
$f_2 = -0.05$ m,
$L = 0.05$ m.

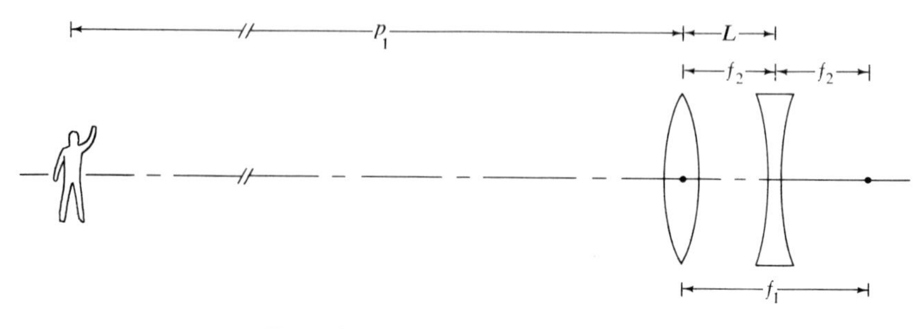

Does the image appear bigger or smaller?

1.7 Find and describe the image in the following mirror combinations:

(a) Simple magnifier:

$Y_{obj} = 1$ mm,
$p = 5$ cm,
$f = 6$ cm.

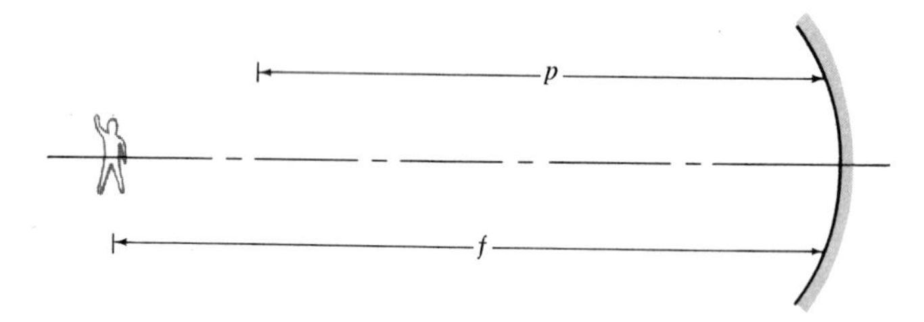

(b) Compound microscope:

$Y_{obj} = 0.01$ mm,
$p_1 = 1.1$ cm,
$f_1 = 1$ cm,
$f_2 = 10$ cm,
$L = 18$ cm.

Comment on possible restrictions on the first mirror.

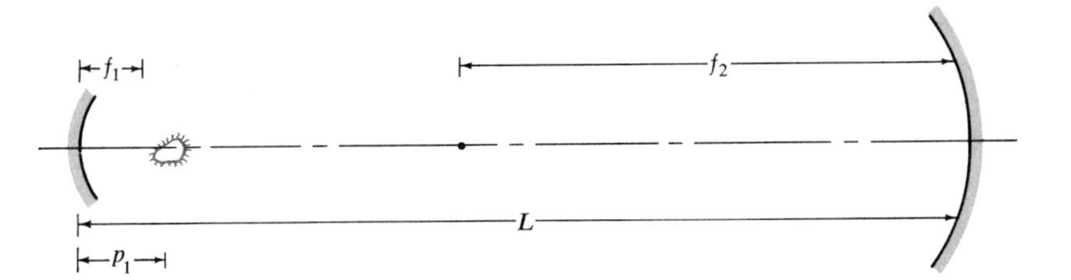

(c) Astronomical telescope:
$Y_{obj} = 2.5$ thousand miles,
$p_1 = 0.25$ million miles,
$f_1 = 10x$ miles,
$f_2 = x$ miles,
$10x <<< 0.25$ million.
Does the image *appear* bigger or smaller? *Hint:* Keep at least the second term in q_1 before approximating. (See Appendix A for help with approximation.)

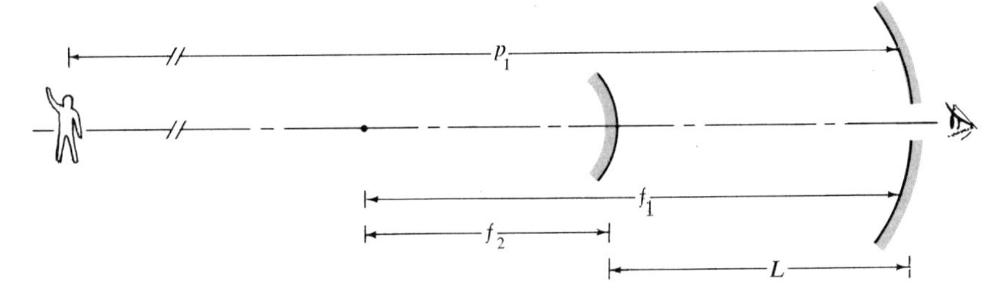

(d) Opera glass:
$Y_{obj} = 2$ m,
$p_1 = 50$ m,
$f_1 = 0.10$ m,
$f_2 = -0.05$ m,
$L = 0.05$ m.
Does the image appear bigger or smaller?

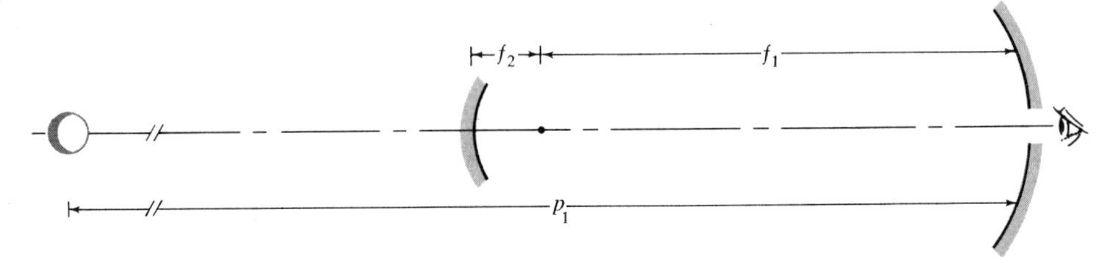

1.8 An ideal camera lens of focal length f and diameter d is used to photograph an object at a distance p in front of the lens. A real image is formed on the film at a distance q behind the lens. The object is a square of area A, each point of which radiates light isotropically. The object radiates a total power P. The intensity at its surface is $I = P/A$.

(a) Find the power P' delivered to the film.

(b) Find the intensity I' in the image formed on the film. Express your answer as a function of I, d, f, and p.

(c) Show that for $10f \leq p \leq \infty$, one makes an error of less than 25 percent by regarding I' as a function only of I and the ratio f/d (the f number of the lens).

1.9 Three children line up for their Christmas photo, as shown. Only the one at distance p is in sharp focus. A point on one of the other children forms not a point but a small "blur" circle of diameter δ on the film. If we keep this circle no bigger than a typical silver grain (say, 1 μm, or 10^{-6} m), then the picture appears correctly focused.

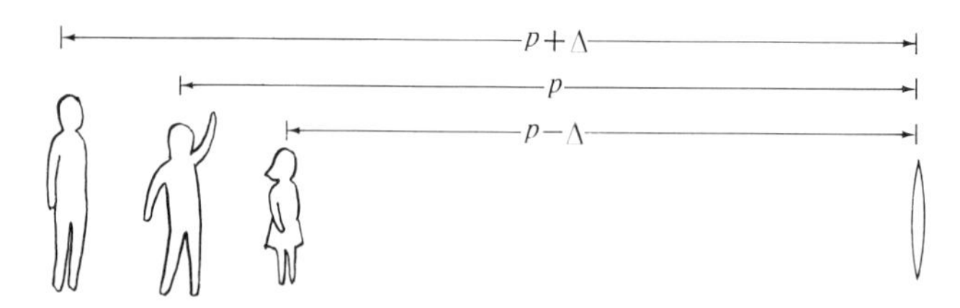

(a) Find δ in terms of Δ, f, p, and the lens diameter d.

(b) If $\Delta/p \ll 1$, express δ in terms of Δ, d, and the object and image sizes.

(c) For best depth of focus on a given subject, what must we do? What does this imply for a "pinhole" camera?

1.10 Find the first three real images, and describe them. P is a plane mirror and C is a curved one of focal length f.

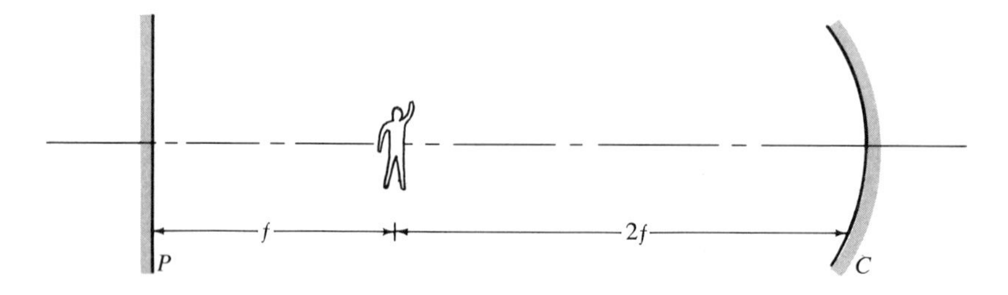

1.11 A goldfish lives in an aquarium full of water with index 4/3. In this water is an air bubble containing a fly. Using the positions shown in the figure, find:

(a) Where the fish appears to be, to an observer outside the tank.
(b) Where the fish appears to be in relation to the fly.
(c) Where the fly appears to be in relation to the fish.
(d) Where the fly appears to be in relation to the outside observer.

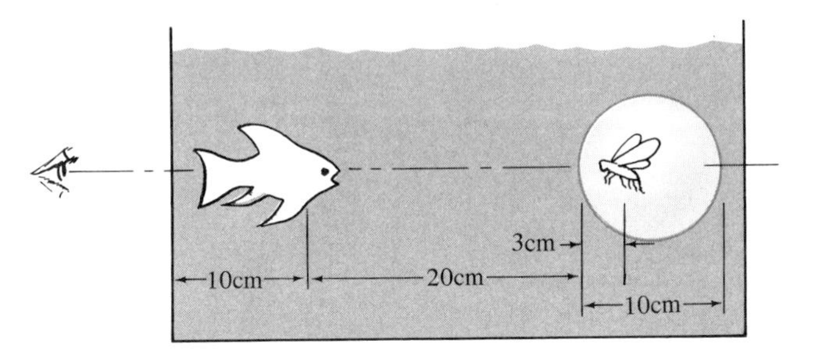

1.12 A hemispherical drop of water rests on a glass slide. A microbe in the water is $R/2$ above the glass, where R is the radius of the drop. The index of water is 4/3 and that of the glass is 5/3. The glass is R thick. Where is the image of the microbe, looking (a) from above? (b) from below?

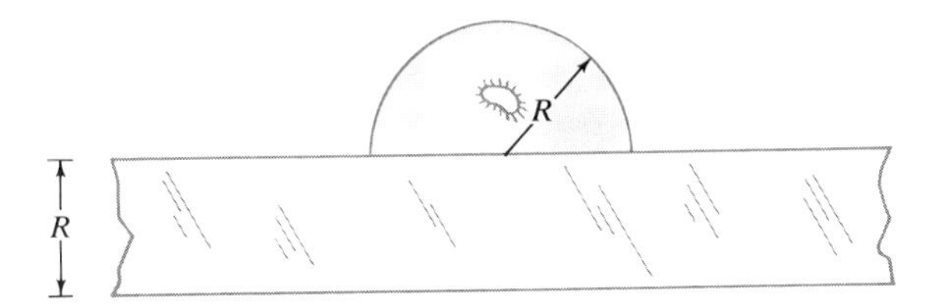

2 Waves: description

Waves are of great importance in most of physics. This chapter will describe them in physical and mathematical terms: physically, they are carriers of energy, momentum, and angular momentum; mathematically, their properties can be specified by a few simple equations. We concentrate at first on mechanical waves, which move through tangible media and are governed by equations derived from the laws of mechanics. Later we will see that the same description applies to electromagnetic waves (including light), which propagate in vacuum and are deduced from Maxwell's equations for the electromagnetic fields. Also in this chapter we introduce a very specific mathematical description, that of the sinusoidal wave. This will enable us to study in detail the behavior of a kind of wave for which the mathematics is not too difficult and which can later be generalized to more complicated forms.

2.1 Physical description

The easiest waves to study are those induced on a "string", since the phenomena in question are simple to demonstrate. Imagine yourself holding one end of a long rope and a partner holding the other, keeping the whole clear of the ground. Pump your end of the rope up and down a few times, and a short train of waves will be generated. These travel down the rope toward your partner, maintaining roughly the shape of the train as it first appeared. This ability to

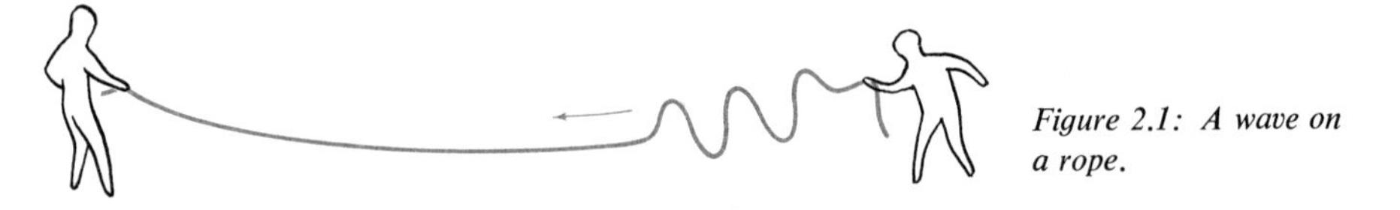

Figure 2.1: A wave on a rope.

travel along the string without changing shape will turn out to be a crucial characteristic of what we will call a wave. Before we try to use such a description, however, let us look at the various phenomena which are familiarly associated with this wave.

First, it takes a certain amount of your energy to get the wave started. When the wave reaches him, your partner is going to feel the rope exerting forces on his hand. If his hand moves, work is done on it; so it is apparent that the energy you put into generating the wave has reached him. Thus, the wave carries *energy*. Since you have moved your partner's hand, you have changed its momentum, so the wave on the rope has also carried *momentum*. Now imagine that you swing your hand in a circle a few times, as if the rope were a jump rope. The circular disturbance runs down the rope, just as did the one in the vertical plane. This circular motion is also a wave and, by the same reasoning as before, it carries both energy and *angular momentum* (since your partner's hand starts to go around in a circle, it has acquired angular momentum). Notice that these three quantities are also the ones which figure in the great conservation laws of physics.

Now let us look again at the thing we are going to call a wave, and watch it as it travels down the rope: Of course, in most real ropes, the wave gets smaller as it progresses; it also changes shape somewhat. We will deal with these phenomena later on, but for the moment let us suppose that we have an ideal rope in which such tendencies are negligibly small. Such a rope can in fact be made, so the idealization is not entirely artificial.

What is the wave, then? In our example, it is a change of position of the rope so that it lies along a new curve. But this new configuration of the rope is one which changes with time and in a very special way. Suppose we take a snapshot photograph of the rope at some early time and another one later. Ahead of the wave and behind it, the rope lies along its undisturbed position, and the two snapshots show an identical geometrical structure in the disturbed part. In fact,

Figure 2.2: Progression of a wave.

if we were to enlarge one picture to life size, we could walk along beside the rope and match the picture with the wave point-for-point as long as we moved at the right speed. We call this speed the *wave speed c*.

2.2 *Mathematical description*

Now how should we summarize this strictly geometrical and pictorial aspect of the wave? The snapshot could be described in analytic geometry as a curve $y(x)$, where y is the distance the rope has moved from its equilibrium position, specified at each point x along the rope. Since the wave moves along the rope, it is also a function of time: $y(x, t)$. The feature which distinguishes this function of space and time from some general one is that successive snapshots [say, $y(x, 6 \text{ sec})$ and $y(x, 7 \text{ sec})$] look identical except for the position of

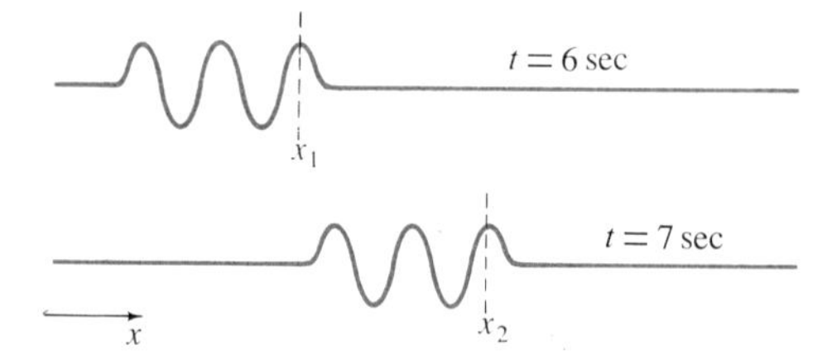

Figure 2.3: Successive snapshots of a wave.

the wave along the rope. The entity we think of as the wave does not change its shape; it merely moves down the rope. We can make the statement that the two snapshots look alike by saying that

$$y(x_1, 6) = y(x_2, 7).$$

Then, if we know how x_1 and x_2 are related, we should be able to predict the appearance of a third snapshot at some later time. We do know how the positions are related because we know that we can walk alongside the rope at speed c, matching the snapshot with the wave point-for-point. This means that the special function which describes the wave is $y(x - ct)$.

In the preceding example, we can see that $y(x - ct)$ is the right function, since insertion of the appropriate value for c (that is, $c = (x_2 - x_1)/1$ sec) gives $x - ct = 7x_1 - 6x_2$ for both pairs of values. This means that the y values are the same, as we assumed:

$$y(x_1, 6 \text{ sec}) = y(7x_1 - 6x_2) \quad \text{and} \quad y(x_2, 7 \text{ sec}) = y(7x_1 - 6x_2).$$

So far, the functional relationship merely tells us that if y is a *function* of $x - ct$ (that is, if y specifies some mathematical expression in which $x - ct$ is treated as a unit), then the way in which the shape of the rope *changes* in time is a special one. We know that the shape just moves down the rope, as if it were unchangeable (like the snapshot), but it moves as a whole.

More formally, we note that the way to specify that something is constant (the shape of the disturbance) is to see that its derivative is equal to zero. If we let the position of the point where we measure y be a function of time (as it is when we walk beside the wave, matching), then we set the time derivative of $y(x, t)$ equal to zero*:

$$\frac{dy}{dt} = 0 = \frac{\partial y}{\partial x}\frac{dx}{dt} + \frac{\partial y}{\partial t} = \frac{\partial y}{\partial(x - ct)}\frac{\partial(x - ct)}{\partial x}\frac{dx}{dt} + \frac{\partial y}{\partial(x - ct)}\frac{\partial(x - ct)}{\partial t}.$$

So,

$$\frac{dy}{dt} = \frac{\partial y}{\partial(x - ct)}\left\{1 \cdot \frac{dx}{dt} - c\right\} = 0.$$

Here, it is not necessary to know $\partial y/\partial(x - ct)$.

We might alternatively use this method to define c as the wave speed. If c is a positive quantity, the argument of y (that is, the quantity $x - ct$) will remain constant if x increases with time. This specifies a wave moving toward $+x$, but it is equally possible for the wave to move in the other direction so that we must allow the wave $y(x + ct)$. Again performing the differentiation, we see that $dx/dt = -c$, specifying formally a wave propagating toward $-x$.

* The symbol $\partial y/\partial x$ is that for the partial derivative. That is, $y(x, t)$ is differentiated with respect to x only, while holding t and any other variables constant. The distinction is an important one, as seen in this example. Also see Appendix A about this.

The function $y(x \pm ct)$ describes the wave on the rope, but we should allow waves in other media. So we use the general form $f(x \pm ct)$, where the function may be the displacement y or as different a quantity as an electric field.

Although $f(x \pm ct)$ describes the wave, there is no "physics" in the description. By this we mean that no general laws of motion or conservation have dictated the behavior of the medium, which has been simply described in this way. Such laws enter through the *wave equation*, which governs the system and for which $f(x \pm ct)$ is a solution. The situation is similar to that in elementary mechanics: Newton's laws govern the behavior of material objects, but the equations of motion describe how they actually move in a particular situation. The wave equation, like Newton's second law, is a differential equation. That is, it states a relationship between one or more derivatives of some function and certain external constants. Thus, $\mathbf{F} = m(d^2\mathbf{r}/dt^2)$ specifies how the state of motion of a material object changes in response to external forces, and the wave equation

$$\frac{\partial^2 f}{\partial x^2} = \frac{1}{c^2}\frac{\partial^2 f}{\partial t^2}$$

specifies how the function f changes in the medium. In both cases the equations are too general for more than formal use unless the values of f or $\mathbf{r}$ are given for some specified time or place. Such values are "boundary values," which enable us to describe actual trajectories or explicit waves. The great usefulness of the wave equation is that its solutions are always of the form $f(x \pm ct)$; hence, whenever we encounter such an equation, we know that waves will result. In Appendix B we derive this equation for the case of the string and show that it follows directly from Newton's second law. The acceleration in the y direction is the term $\partial^2 y/\partial t^2$, and it is shown that $\partial^2 y/\partial x^2$ represents the curvature, or change of slope of the string. This is the required unbalanced force. Appendix B also contains a derivation for sound waves, in which case f represents the pressure in the medium (for instance, air) through which the wave travels. The important subject of the occurrence of wave equations is outside the scope of this book. Instead, we concentrate on the solutions to the equations (mathematically the same for all wave equations) and the physical phenomena which those solutions describe. The fact that the same solutions occur in widely different fields of physics gives this subject extra interest. In fact, all modern quantum mechanics uses just such solutions and mathematical techniques as we discuss here.

We now come to explicit forms of the function $f(x \pm ct)$. Both $f(x - ct)$ and $g(x - ct)$ are solutions to the wave equation, and their sum is a solution also. Indeed, any linear combination* of solutions is a solution, so that $f(x + ct) \cdot A + g(x - ct) \cdot B + h(x + ct) \cdot C$ is a wave too. This statement, which is easy to prove, forms the basis for the principle of *superposition*, which we will use extensively in later chapters.

What sort of function is f likely to be? The simplest is $f = x - ct =$ constant. This describes a particle in uniform motion. That is, a marble rolling north is a wave if it is not accelerated, according to our description. Such a trivial case is better described without waves. In fact a material particle can be described mathematically as a wave in a more profound sense, which is useful when the particle is one of the very small ones encountered in modern physics. The relationship is not so simple as that implied here, and the wave equation is more subtle than Newton's second law, which governs the marble.

A second function which we might encounter is $y = e^{-\alpha(x-ct)^2}$, which is a possible wave on a string (known as a Gaussian pulse, since it is a single wave pulse in the shape of a Normal, or Gaussian, curve). Such a wave has interesting properties, and is important in a number of areas of physics and technology. However, we will primarily concern ourselves with a simpler wave, the sinusoidal one. (2.1–2.3)

2.3 Sine wave

A sinusoidal wave is so named because the function f has the form $f = A \sin[b(x \pm ct)]$. Notice that the sine here is a mathematical *function*. We will profit by its other identity as a ratio of sides in a right triangle, but its primary role is that of function. That is, $\sin(q)$ is a quantity which takes on a range of values (from -1 to $+1$) as q takes on its possible values. We think of q as an angle only incidentally and then always express it in radians. For instance, when $q = \pi/2$, $\sin(q) = 1$.

The sinusoidal wave is a repeating structure, one *cycle* being

* A linear combination means a sum of terms linear in the components. No products of two components or higher powers of components occur. $A + B$ is a linear combination, but AB and $A^2 + B$ are not.

exactly like every other. We think of it as extending to infinity in both space directions and into the past and future. The fact that this is never physically possible has interesting consequences, to be studied in due course. The symmetry of waves with respect to the two variables x and t makes the snapshot view (time held constant,

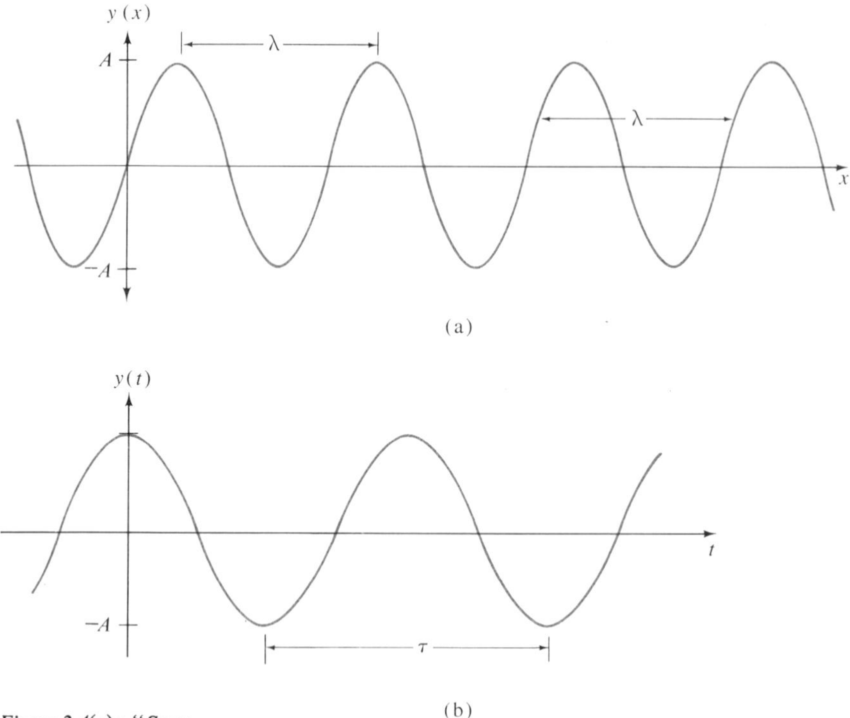

Figure 2.4(a): "Snapshot" of the function $y(x, t)$. This is a graph of $y(x, t_0)$, where t_0 is a constant (fixed) value of t.

(b): "Local" chart of the values of $y(x, t)$ measured by a stationary observer as the wave passes him. Observer is at x_0, a constant (fixed) value of x, so the graph is of $y(x_0, t)$.

y a function of x) appear the same as the local view (y a function of time, at some chosen place). These are shown in Figure 2.4.

Consider now the quantities A and b. A is the *amplitude* of the wave. That is, $\sin[b(x - ct)]$ runs from -1 to $+1$, so y runs from $-A$ to $+A$. For a string, the displacement y is a length; so the amplitude will have the dimension of length in this case. The argument of a function, on the other hand, must be dimensionless; therefore b has the dimension of 1/length. Conventionally, we set it equal to $2\pi/\lambda$, where λ is the *wavelength* of the wave, which is the distance along the wave between repetitions of the same configuration. Stated more rigorously, λ is the distance between points having identical displacements y and slopes $\partial y/\partial x$. Pictorially, it is the distance from crest to crest.

The corresponding temporal quantity is the *period* τ. This is the time between repetitions of a value of y and $\partial y/\partial t$ as measured at one place. It is most easily seen on the graph of $y(t)$. It is more usual to speak in terms of the *frequency* ν with which these values recur, where $\nu = 1/\tau$. Thus, if y reaches its maximum value (crest) every $\frac{1}{7}$ sec, the frequency of the repetition is 7 times per second. We speak of this as 7 cycles per second, expressed as 7 hertz (Hz). (Earlier notation for this which the student may encounter is 7 cps or 7 c/s. Also be careful not to confuse ν—the Greek letter "nu"—with v, which we use for velocity.)

The symmetry of λ and τ is best displayed by writing the function as

$$y = A \sin 2\pi\left(\frac{x}{\lambda} - \frac{t}{\tau}\right).$$

This also points up the very useful relationship $\tau = \lambda/c$, or $\lambda\nu = c$. This former is more than a notational convenience, since the statement $\lambda = \tau c$ is a statement that the wave, traveling at speed c, goes a distance λ in a time τ.

Two further notational modifications are those which eliminate the factor 2π. We can say $\omega = 2\pi\nu$ and $k = 2\pi/\lambda$. These are most convenient in mathematical manipulations, so we include them here only for completeness. In our more concrete applications of sinusoidal waves, the form using λ and ν will be more useful.

The wave described above requires that $y = 0$ at $x = 0$ when $t = 0$. This restriction may be removed by the inclusion of a *phase factor* ϕ, which may be thought of as specifying the relative positions of the point chosen as $x = 0$ and the point at which the curve

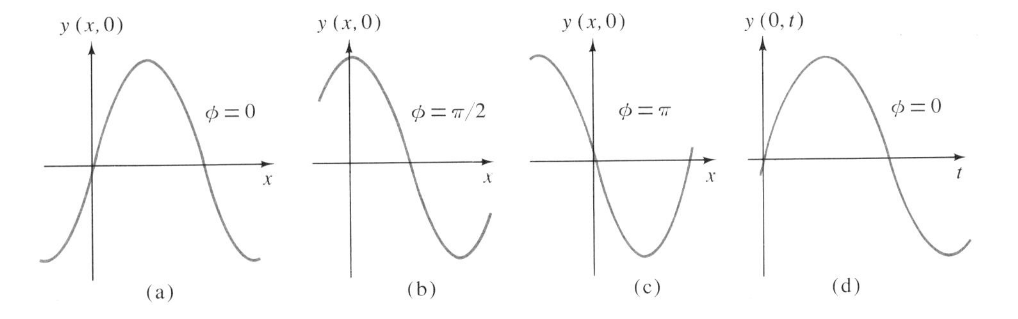

Figure 2.5: The wave near x, or t = 0, for different phase constants.

$y = A\sin[(2\pi/\lambda)(x - 0)]$ crosses the x axis. This is illustrated in Figure 2.5. The general form is (2.4–2.5):

$$y(x, t) = A \sin\left[\frac{2\pi}{\lambda}(x \pm ct) + \phi\right].$$

When $\phi = 0$, we have $y = A\sin\left[\frac{2\pi}{\lambda}(x \pm ct)\right]$, and when $\phi = \pi$, we have $y = -A\sin\left[\frac{2\pi}{\lambda}(x \pm ct)\right]$. Also, when $\phi = \pi/2$, the function is simplified by using its trigonometric character to make the identification

$$y = A \sin\left[\frac{2\pi}{\lambda}(x \pm ct) + \frac{\pi}{2}\right] = A\cos\left[\frac{2\pi}{\lambda}(x \pm ct)\right].$$

The function $y(x, t)$ describes mathematically the entity which we recognize as a wave. Let us see if we can extract from it some of the physical quantities which we might expect to measure in a laboratory. Consider as an example a wave on the surface of the sea. A scale is painted on a piling driven into the sea bottom so that we can measure the height of the water at that point for any time. A graph of our measurement looks like Figure 2.6.

Figure 2.6: "Local" picture for the example.

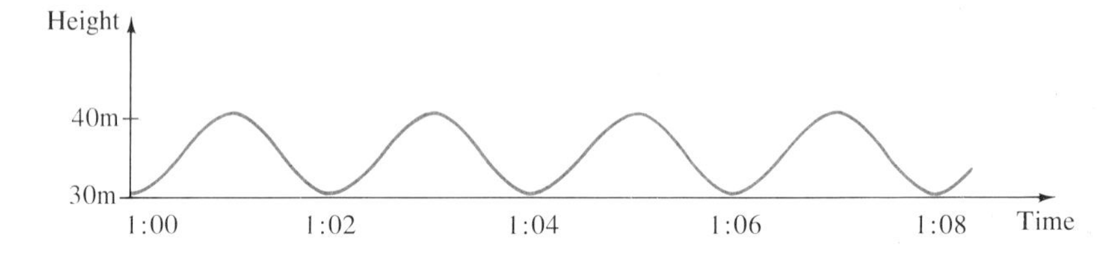

Suppose that at 1:03 P.M. we take a snapshot of the water surface, and it looks like Figure 2.7. Then the four parameters of our equation are easily obtained. First, the amplitude is 5 m. The height of the surface moves between 30 and 40 m, and this total excursion

Figure 2.7: "Snapshot" for the example.

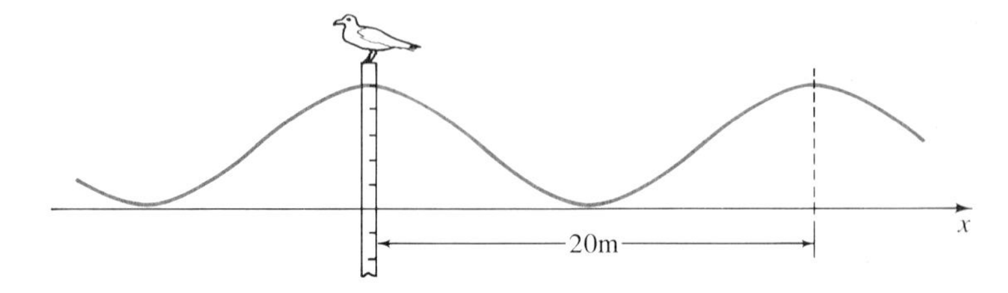

is $2A$. We measure y from the equilibrium point, at 35 m. Next we may ask for the wavelength. Our snapshot shows this directly as 20 m. The value of c is not so directly evident, but we can obtain it from λ and τ. The period is measured on the first graph as $\tau = 2$ min or, more conventionally, $\tau = 120$ sec. Now $c = \lambda/\tau$, and so our value of c is $c = \frac{1}{6}$ m/sec. The remaining parameter is ϕ. For this we must choose a zero point for time and position. Let us say $x = 0$ at the piling, and $t = 0$ at 1:00 P.M. We know the value of y at this time and place, so we put the known values of x and t in the argument to obtain

$$y(0, 0) = +A \sin\left[\frac{2\pi}{\lambda}(0 - c0) + \phi\right] = -A.$$

Hence, $\sin(\phi) = -1$, or $\phi = -\pi/2$.

So far we have found $A = 5$ m, $\lambda = 20$ m, $\tau = 120$ sec ($\nu = \frac{1}{120}$ Hz), $c = \frac{1}{6}$ m/sec, and $\phi = -\pi/2$. Now, what are the other physical quantities involved? A buoy on the water surface is riding up and down. If we know where it is (that is, at what value of x, say, x_0), then we know how high it is at each instant, since this is given by $y(x_0, t)$, which we know. For instance, if the buoy is 2.5 m away ($x_0 = 2.5$ m), its height at 1:35.5 P.M. is

$$y = \left\{5 \sin\left[2\pi\left(\frac{2.5}{20} - \frac{13.5 \cdot 60}{6}\right) - \frac{\pi}{2}\right] + 35\right\} \text{ m}$$

or

$$y = \frac{5}{\sqrt{2}} \text{ m} + 35 \text{ m}.$$

2.4 Momentum and energy

In the preceding example, we may also ask how fast the buoy is moving and in what direction. It is moving up and down, of course, so the velocity in question is $v_y = \partial y/\partial t$. Note that this is *not* the wave velocity. The vector $\mathbf{v}_y$ is vertical, at right angles to the direction in which the wave is moving. Performing the differentiation, we see that

$$v_y = -\frac{2\pi}{\tau} A \cos\left[2\pi\left(\frac{x}{\lambda} - \frac{t}{\tau}\right) + \phi\right].$$

If the buoy is at $x_0 = 10$ m, then at 30 sec after 1:00, its velocity is

$-\pi/12$ (m/sec), the negative sign meaning that it is moving downward. Similarly, we find its acceleration to be

$$a_y = \frac{\partial^2 y}{\partial t^2} = -\left(\frac{2\pi}{\tau}\right)^2 A \sin\left[2\pi\left(\frac{x}{\lambda} - \frac{t}{\tau}\right) + \phi\right].$$

The net force on the buoy is its acceleration times its mass. This net force comes completely from the wave, since gravity is balanced by the buoyant force of the water whether the wave is present or not. The water itself moves up and down and is acted on by unbalanced forces.

$$\mathbf{F}_{\text{net}} = m_{\text{buoy}}\,\mathbf{a} \qquad \text{vertically.}$$

Energy is required to accelerate the buoy, and at any given time the buoy has momentum. We can calculate these also. The momentum is

$$p_{\text{buoy}} = m_b v_y = m_b 2\pi\nu A \cos\left[2\pi\left(\frac{x_0}{\lambda} - \frac{t}{\tau}\right) + \phi\right].$$

This has a maximum value of $2\pi\nu m_b A$ when the buoy is at its equilibrium position. At that position, its kinetic energy is also a maximum:

$$(\text{KE})_{\text{buoy}} = \frac{1}{2} m_b v_y^2 = \frac{m_b}{2}\left(\frac{2\pi}{\tau} A \cos\left[2\pi\left(\frac{x}{\lambda} - \frac{t}{\tau}\right) + \phi\right]\right)^2$$

$$= \frac{m_b}{2}\left(\frac{2\pi}{\tau}\right)^2 A^2 \cdot 1 = 2\pi^2 m_b \nu^2 A^2.$$

When the buoy is not at the equilibrium position, it has gravitational potential energy:

$$(\text{PE})_{\text{buoy}} = m_b g y = m_b g A \sin\left[2\pi\left(\frac{x}{\lambda} - \frac{t}{\tau}\right) + \phi\right].$$

These results depend on the mass of the buoy, but a drop of water near the surface is treated the same way and would enable us to find the total energy in the wave. We do the same calculation for the wave on a string, since it is for this that we know the mechanical details.

Taking ρ as the linear density a small segment of the string, of length Δx, has a mass $\rho\,\Delta x$ and velocity

$$v_y = -\frac{2\pi}{\tau} A \cos\left[\frac{2\pi}{\lambda}(x - ct) + \phi\right].$$

The momentum of this length of the string is thus

$$p\,\Delta x = (\rho\,\Delta x)v_y = -\rho \cdot 2\pi\nu A \cos\left[\frac{2\pi}{\lambda}(x - ct) + \phi\right]\Delta x,$$

so that p is the momentum per unit length, or the momentum density. The kinetic-energy density is

$$\mathscr{E}_k = \frac{1}{2}\rho(2\pi\nu A)^2 \cos^2\left[\frac{2\pi}{\lambda}(x - ct) + \phi\right] = \mathscr{E}_0 \cos^2\left[\frac{2\pi}{\lambda}(x - ct) + \phi\right].$$

The potential energy of the segment turns out to be

$$\mathscr{E}_p\,\Delta x = \mathscr{E}_0 \sin^2\left[\frac{2\pi}{\lambda}(x - ct) + \phi\right]\Delta x,$$

as can be found by calculating the work necessary to move the mass $\rho\,\Delta x$ a distance y against the force $(\rho\,\Delta x)a$ (where a is the acceleration). Thus the total energy density is $\mathscr{E} = \mathscr{E}_0$, a constant. The quantities p, $\mathscr{E}_k$ and $\mathscr{E}_p$ are all functions of $x - ct$ and therefore are waves. Notice that the average value of p with respect to either x or t is zero. This means that there is no *net* motion of the string because as many points move up as move down. This is one way of saying that the average value of p with respect to x is zero. Similarly, any given point spends just as much time moving up as it spends in moving down, so the average with respect to time is zero. That the same is not true of the kinetic or potential energy can be seen a number of ways: The area under the curve,

$$\cos^2\left[\frac{2\pi}{\lambda}(x - ct) + \phi\right],$$

is greater than zero, as can easily be seen by plotting it out. Appendix A shows formally that the average equals $\frac{1}{2}$, and this can be remembered by observing that 1 is the average of $\sin^2 q + \cos^2 q = 1$, and that the function in question is one-half of this value. So the *average* kinetic energy per unit length is $\langle\mathscr{E}_k\rangle_{av} = \rho\pi^2\nu^2A^2$. It will be true in general that the average energy carried by the wave is proportional to the *square* of the amplitude, regardless of what wave we are calculating. (2.6–2.7)

We have seen that waves are nonequilibrium configurations of a medium, and that they propagate in that medium according to the mathematical description $f(x, t) = f(x \pm ct) = g(x + ct) + h(x - ct)$. They transport energy, linear momentum, and angular momentum.

The way in which the laws of physics control the behavior of the wave is described by the wave equation:

$$\frac{\partial^2 f}{\partial x^2} = \frac{1}{c^2}\frac{\partial^2 f}{\partial t^2}.$$

A consequence of this is that all solutions of the form $f(x, t)$ above describe possible waves. By means of *boundary conditions*, we select the one actually required in a given physical situation. Two such statements as to what the function f is at a given time and place suffice to specify the wave entirely. [Two, mathematically, because the wave equation is second order, which in turn allows the two solutions $g(x + ct)$ and $h(x - ct)$]. Finally, we note that the sum of two solutions is also a solution. This is the mathematical statement of the principle of *superposition*, which will be the subject of most of our later investigations. First, when we discuss reflection, we will show how it may be used to simplify the use of boundary conditions. Then we will use it to discuss the phenomenon of standing waves, a particular consequence of reflections. In these cases only the sign of c is different in the two solutions we add.

In Chapter 3 we will also superpose two periodic waves with different frequencies to obtain beats, and we will look briefly at the use of this principle to construct nonsinusoidal periodic waves and even nonperiodic ones. Later we will superpose two or more waves of different phase to obtain the phenomena of interference and diffraction.

EXERCISES

1. In Figure 2.3, let $x_1 = 5$ m, $x_2 = 7$ m. At what time is $y(17\text{ m}, t) = y(x_1, 6\text{ sec})$?

2. Prove explicitly that $y(x, t) = \exp[-b(x + ct)^n]$ is a solution of

$$\frac{\partial^2 y}{\partial t^2} = c^2\frac{\partial^2 y}{\partial x^2}.$$

3. Show that $y(x, t) = \sum_{n=0}^{\infty}[b(x - ct)]^{5n+1}$ is a solution to

$$\frac{\partial^2 y}{\partial t^2} = c^2\frac{\partial^2 y}{\partial x^2}.$$

4. Show that

$$\frac{\partial}{\partial t}\sin\left(\frac{2\pi x}{\lambda} - 2\pi\nu t + \phi\right) = -c\frac{\partial}{\partial x}\sin\left(\frac{2\pi x}{\lambda} - 2\pi\nu t + \phi\right).$$

5. The light intensity curve for a cloudless sky, $I(t)$, is approximately a sine wave. At the equinox (equal light and dark periods), what is the appropriate phase constant if the zero of time is midnight? What modifications should be introduced to compensate variation of position on the earth?

6. Find the total energy stored in a violin string vibrating with amplitude 1 cm and frequency 440 Hz, averaged over 1 cycle, the first tenth of a cycle, and 1000.1 cycles.

PROBLEMS

2.1 A traveling wave is described by the equation

$$y = \exp[(-az^2 - bt^2 - 2\sqrt{ab}\, zt)].$$

(a) In what direction is the wave traveling?
(b) What is the wave speed?
(c) Sketch this wave for time $t = 0$ and for time $t = 3$ sec, both using $a = 144/\text{cm}^2$, $b = 9/\text{sec}^2$.

2.2 A wave is described by the equation

$$y = \frac{\sin^2\{za[1 + q(t/z)]\}}{(z + qt)^2}$$

(a) In which direction does the wave travel?
(b) What is the wave speed?
(c) Sketch the graph $y(8\text{ m}, t)$ from $t = -10$ sec to $t = +10$ sec, for $q = 8$ m/sec, $a = \frac{1}{8}\text{ m}^{-1}$.

2.3 The leading edge of a wave pulse on a string has a slope of 0.010. The wave speed is 5.0 m/sec. What is the transverse particle speed (v_y) at the leading edge of the pulse?

2.4 The accompanying illustration gives the information about a sinusoidal wave. Write the equation of the wave. $x = 9$ m, $c = 5$ m/sec.

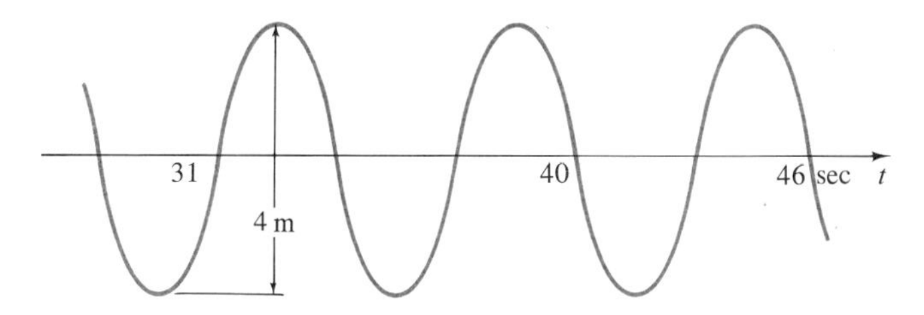

2.5 The wave shown below travels on a string of linear density 3 g/cm.

(a) Find the energy and momentum of the part of the string between $x = 40.99$ cm and $x = 41.01$ cm at time $t = 29$ sec.

(b) Find the force on this part, and draw a graph of it.

(c) What is the tension in the string? (See Appendix B.) $c = 7$ m/sec, $t = 25$ sec.

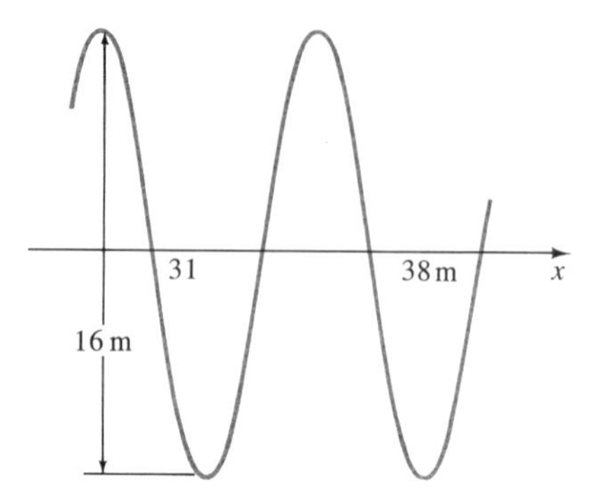

2.6 A massless rubber band is stretched and weighted every 2 cm with 10-g beads. We observe the indicated configurations at $t = 2$ sec and at $t = 4$ sec. (Remember that the beads move *only* in the y direction.)

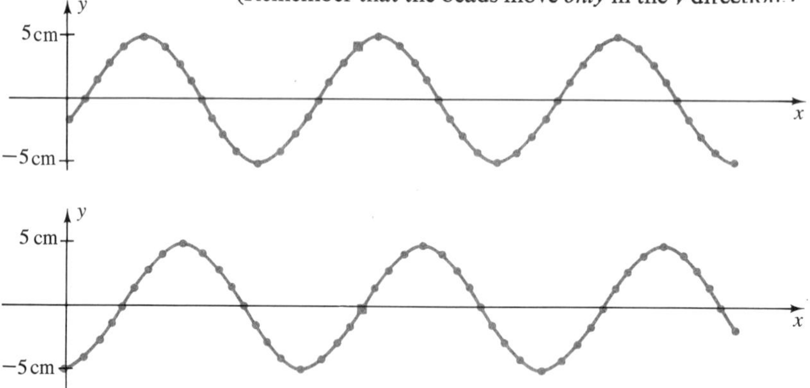

(a) Write the equation of the wave. Assume that less than 1 cycle has passed between pictures.

(b) Find the energy and momentum of the square bead when $t = 12$ sec.

(c) Find the force on the square bead as a function of time, and draw a graph of this.

(d) What is the tension in the rubber band? (See Appendix B.)

2.7 A wave of arbitrary shape occupies a 1-m length of a string. Ahead and behind, the string lies along $y = 0$. Show that the net momentum carried by the wave is zero.

3

Superposition: reflection, standing waves, group velocity

In this chapter we start to deal with problems involving the superposition of two or more waves. If one wave raises a piece of string 5 cm and another lowers it 2 cm, then its net displacement is +3 cm when both waves are present. This simple addition of the waves is what we mean by superposition. More generally, we can describe it as the addition of two or more waves to give the total wave disturbance

$$y_T(x, t) = y_1(x, t) + y_2(x, t).$$

Addition of the two waves for some given value of x and t is trivial, but to add two analytic expressions of some general functional form will be more difficult.

3.1 *Superposition*

Superposition is an important concept because it tells us some physical facts about the medium through which the wave passes. There are many real situations in which waves do not superpose. An important one is the case of a nonlinear medium. A string stretched beyond its elastic limit is such a medium. A wave on the string might break it or deform it permanently, losing energy in the process.

Another kind of nonlinear string is one confined to a pipe. If the wave amplitude exceeds the radius of the pipe, no further displacement is possible.

Waves describing the common vector quantities such as force, displacement, or electric field will superpose if the medium is ideal and linear. But waves of energy, for instance, do not. Think of the kinetic energy of the string in Chapter 2: at x_0, t_0,

$$v_T = v_1 + v_2 \qquad \text{but} \qquad \mathscr{E}_T = \frac{\Delta x(v_1 + v_2)^2}{2} \neq \mathscr{E}_1 + \mathscr{E}_2,$$

although both v and $\mathscr{E}$ obey the wave equation. It is precisely this difference that will concern us extensively in later chapters, since light intensity is proportional to the square of the electromagnetic field amplitude, just as kinetic energy is proportional to the square of the transverse velocity on the string. It is also this difference which plays a significant role in quantum mechanics, since the waves which describe material bodies have superposing amplitudes but nonsuperposing "intensities." There, as with light, it is the nonsuperposing quantity which is measurable.

Superposition, then, describes the behavior of quantities which normally add as vectors in ideal media (and thereby defines "ideal" media). Figure 3.1 shows two pulses on a string. At every point,

$$y_T(x, t) = y_1(x, t) + y_2(x, t)$$

so that the pulses appear to ignore and pass through each other. (3.1–3.3)

3.2 *Reflection*

We can use superposition to simplify the description of a wave undergoing reflection. First look at the physical details of such a reflection.

Consider a wave pulse traveling along a string. We can describe it as $f(x - ct)$ only as long as the string is infinitely long. But if the string has an end (and most do), we must take this into account. For instance, we may fix the string to a wall so that point B cannot move. At a time well before the pulse encounters the wall, its leading edge accelerates the string (upward in the figure). This happens because, as Newton's second law tells us, unbalanced forces on a section of the string result in its being accelerated. But the wall is too massive to move when accelerated, so it must exert a force back

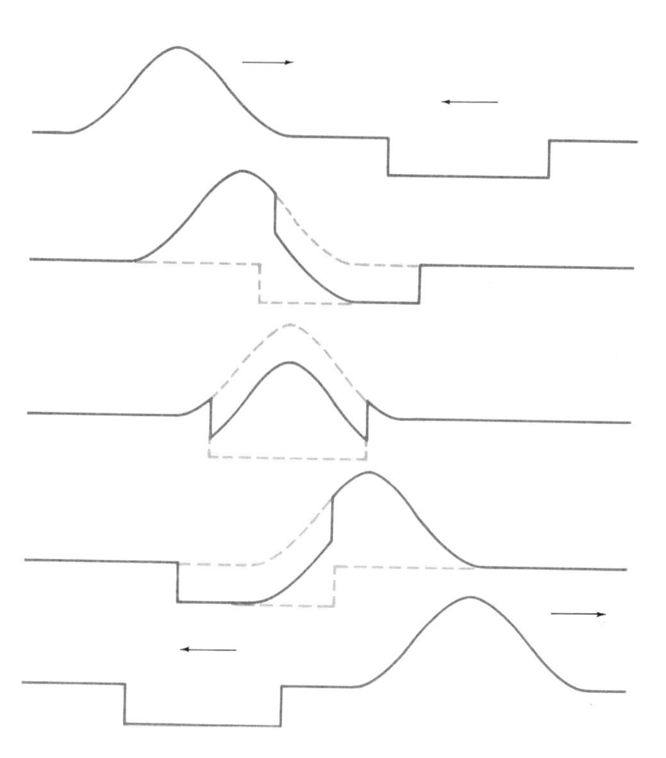

Figure 3.1: Superposition of two wave pulses.

on the string according to Newton's third law. Since this force is equal in magnitude and opposite in direction to the one which the string exerts on the wall, it is just the force which would be exerted by the leading edge of an inverted pulse coming from an (imaginary) string extending past the wall. This is the nub of the method of images, which we will use shortly. The forces exerted on point B by the wall, then, are opposite in direction to those exerted by the rest of

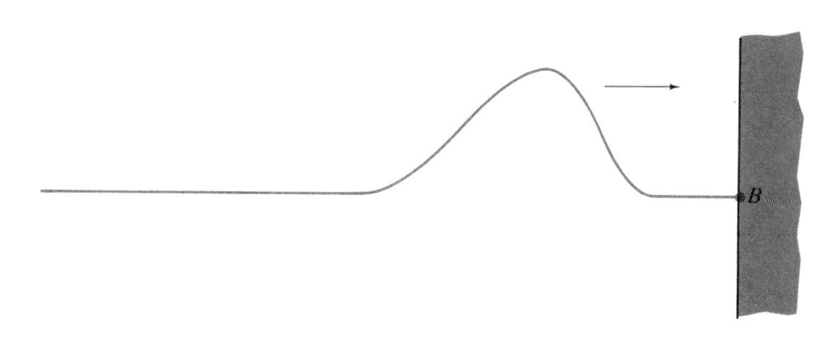

Figure 3.2: Pulse on a string with one end fixed.

the string, and are sequenced so that the wave they generate moves away from B, back along the string, and can be described as $-f(-x - ct)$. The reason the forces of the third law do not generate backward waves all along the string is that the string "gives" everywhere else.

In order to understand this as thoroughly as we should, we might follow a simple pulse and apply Newton's laws as we go. This is done in detail in Appendix C. It is all rather tedious, and in fact quite difficult for most waves. A simpler procedure is to identify the incident wave with $f(x - ct)$ and the reflected one with $-f(-x - ct)$, both of which describe possible waves on the string. We then apply the boundary condition (physical fact) that point B never moves. To do this, we make the following argument: What goes on to the right of the wall does not concern us; therefore we can *imagine* any configuration we like for the string, which we pretend extends to the right of the wall. The only restriction is that any interaction of the imagined wave (to the right) with the real string on the left must agree with our observations on the left. This all results in saying that the total disturbance consists of $f_T(x, t) = f((x - x_B) - ct) - f((-x + x_B) - ct)$. This is a superposition of a wave to the left and one to the right, symmetrically placed around the point B (where $x = x_B$). For *all* times, the value of f_T at B is zero. This is one

Figure 3.3: Reflection of a wave pulse.

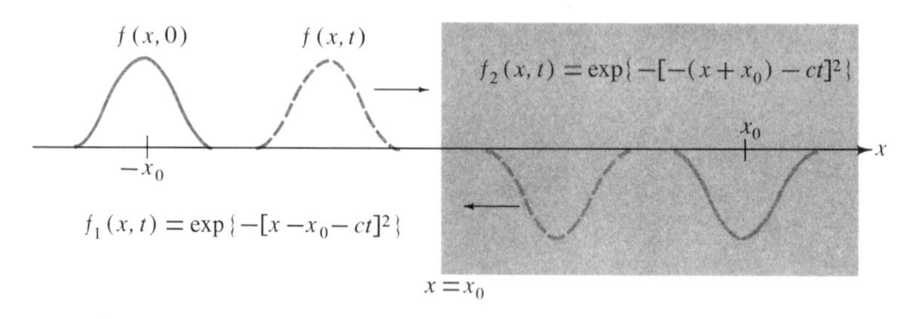

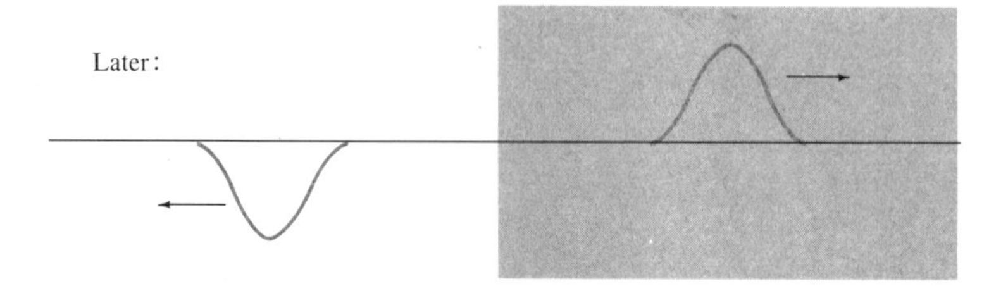

of the results we want. Then, if we are given an initial value for f_T, say $f_T(x, t_0)$, we have completely specified the wave.

For instance, suppose a snapshot at $t = 0$ shows a pulse of the form $\exp[-(x - x_0)^2]$.* A second snap at time t shows that the pulse has the form $\exp\{-[(x - x_0) - ct]^2\}$. Then the boundary at $x = x_0$ is taken into account by the general form $\exp\{-[(x - x_0) - ct]^2\} - \exp\{-[(x_0 - x) - ct]^2\}$. This is shown in Figure 3.3.

We may well ask whether the solution given by this apparent trick is the only one possible. Is it, in fact, unique among the infinity of solutions to the wave equation just because it matches up on the boundary? The answer to this question is "yes." The mathematical theorem which states this is the Uniqueness theorem, and is very important in the solution of boundary value problems. In this special case, where we know the form of the wave when there is no boundary, only the solution we have given satisfies the boundary condition for all time. The student may be familiar with image problems in electrostatics, which are treated by a similar technique and for which the Uniqueness theorem is usually written.

The behavior of the string with a free end is given as a problem at the end of this chapter. We leave for Appendix C the more general problem of two connected strings with different wave speeds. Here, we consider the situation in which an infinite sinusoidal wave undergoes reflection at a wall.

3.3 *Standing waves*

Let the incident wave be $f(x, t) = A \sin[(2\pi/\lambda)(-x - ct)]$. Putting a fixed wall at $x = 0$, the complete wave must be

$$F(x, t) = A \sin \frac{2\pi}{\lambda}(x - ct) - A \sin \frac{2\pi}{\lambda}(-x - ct)$$

$$= A\left\{\sin \frac{2\pi}{\lambda}(x - ct) + \sin \frac{2\pi}{\lambda}(x + ct)\right\}.$$

This sum of two sinusoidal functions is the first case we have encountered of the superposition of two real, physical waves which exist simultaneously everywhere on the string. Since we will see a lot of this sort of thing, it is worth some care in the first encounter. There are at least three useful ways of performing the addition. The

* $\exp\{x\}$ means e^x.

first and most obvious way is to add graphically, point by point. This is done in Figure 3.4. It is clearly not the most convenient way of

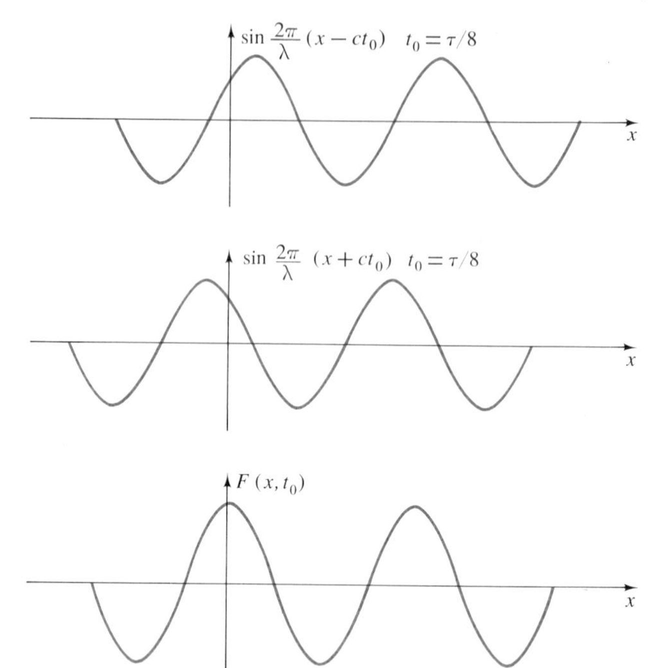

Figure 3.4: Point-by-point addition of two sine waves.

proceeding. The next method is to use the trigonometric aspect of the functions to add them by means of the trigonometric identities. Thus

$$\sin\alpha + \sin\beta = 2\sin\left(\frac{\alpha+\beta}{2}\right)\cos\left(\frac{\alpha-\beta}{2}\right);$$

so

$$\sin\left[\frac{2\pi}{\lambda}(x-ct)\right] + \sin\left[\frac{2\pi}{\lambda}(x+ct)\right] = 2\sin\left(\frac{2\pi}{\lambda}x\right)\cos\left[\frac{2\pi}{\lambda}(-ct)\right]$$

$$= 2\sin\left(\frac{2\pi}{\lambda}x\right)\cos(2\pi\nu t).$$

Now take the usual snapshot of this to see $y(x, t_1)$, or measure y at some point x_1 in order to find $y(x_1, t)$, as shown in Figure 3.5.

Notice that y is zero along the whole string at some time ($t = (2n + 1)\tau/4$, where n is any integer). Similarly, y is always zero

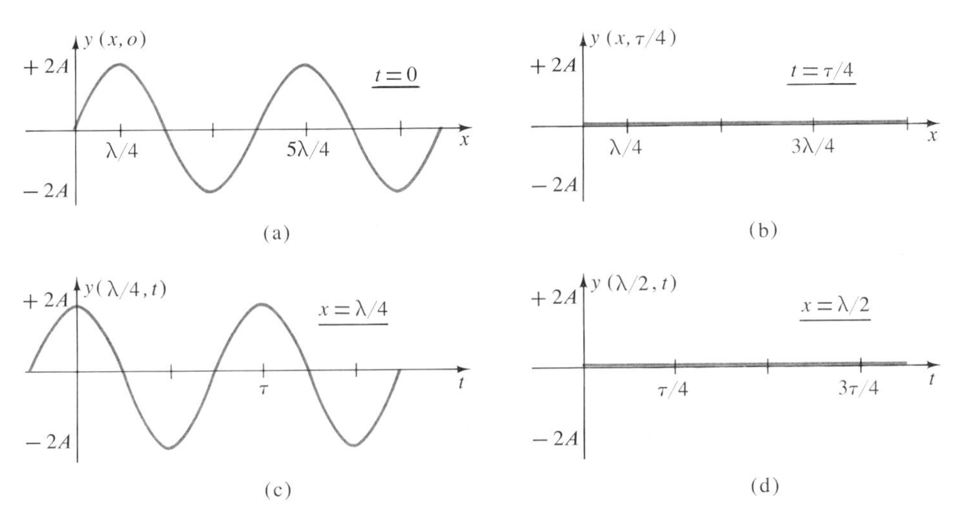

Figure 3.5: Standing wave, various views.

at some points ($x = (2m + 1)\lambda/4$, where m is an integer). These points are called *nodes* of the standing wave. The vigorously moving points halfway between them are called *antinodes*. The term "standing wave" is applied here, of course, precisely because the nodal and antinodal points seem to stand still rather than to move along x. Although there are two traveling waves present, there is no appearance of motion to the right or left.

3.4 Phasors

The third way of adding sinusoidal functions is known as the *method of phasors*. This is convenient in a limited number of situations, but is most useful when a qualitative understanding will suffice. Here we again use the trigonometric aspect of the functions, this time thinking of $A \sin \alpha$ as one component of a vector of magnitude A, at an angle α to some horizontal axis. We say *some* axis, since it is not the x axis. This vector is *not* a vector in real physical space; rather it exists in an arbitrary space which we invent for its benefit. Let us label the axes in this space R and I so as not to confuse them with x and y. If we take α as the angle $\mathbf{A}$ makes with the R axis,

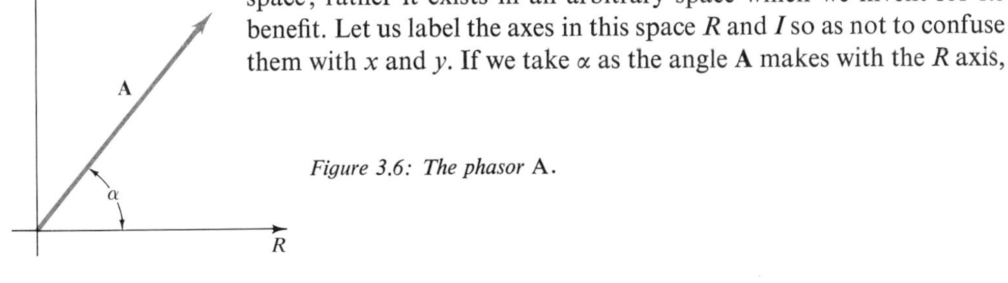

Figure 3.6: The phasor **A**.

then $A \sin \alpha$ is the I component of $\mathbf{A}$. To add the two functions $A \sin \alpha$ and $B \sin \beta$, we simply add $\mathbf{A}$ and $\mathbf{B}$ as vectors and take the I component of the resultant, as shown in Figure 3.7. This indicates

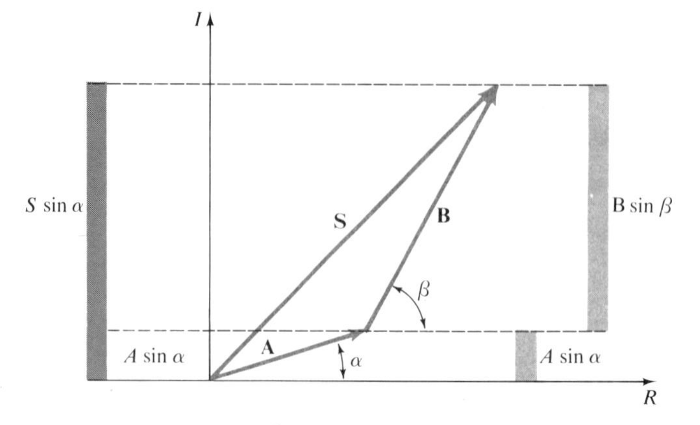

Figure 3.7: Addition of two phasors.

pictorially that $A \sin \alpha + B \sin \beta = S \sin \gamma$. Commonly, α, β, and γ will be functions of x and t. Suppose each of these is proportional to t, as our arguments usually are. Then, because each of the angles is increasing as time goes on, each phasor *rotates*. Often we will work with situations in which $\alpha = qt$ and $\beta = qt + \phi$. This means that the *relative* phases of the component vectors will remain constant, which is the reason these vectors have the special name of "phasor." It is, in fact, generally *only* the relative phases which interest us, so these diagrams can often aid us considerably even though the axes are oriented arbitrarily. That is, if we want, we may set $\alpha = 0$, since it is only the difference $\beta - \alpha$ which concerns us.

For the standing wave of interest here, the phasor diagrams are shown in Figure 3.8. Note how easy it is to obtain the qualitative

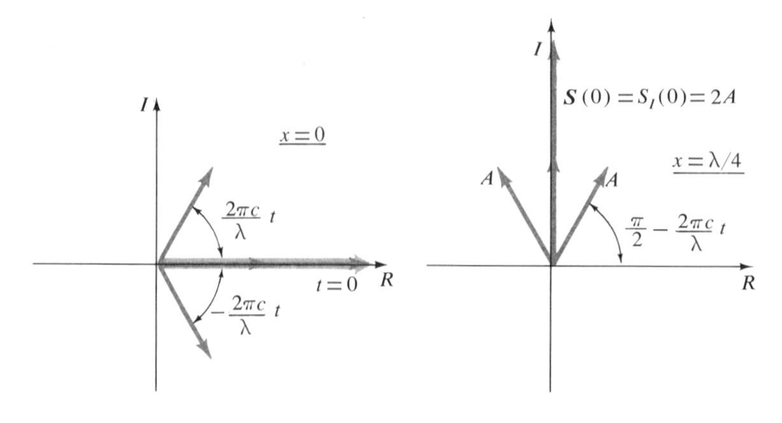

Figure 3.8: Standing waves, phasors. At $x = 0$, $S_I = 0$, for all time. At $x = \lambda/4$, S_I is a sinusoidal function of t, between $+2A$ and $-2A$.

features. Actual calculation, however, is likely to be identical to that of the previous method. Figure 3.9 shows that the angles can also be made functions of x.

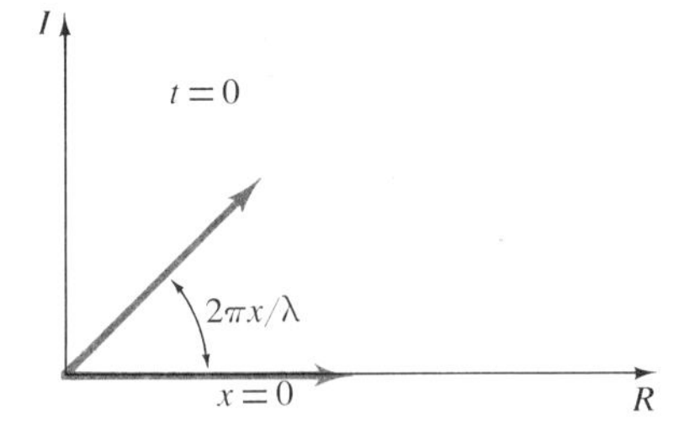

Figure 3.9: A phasor whose argument is a function of x.

Here, S_I is a sinusoidal function of x. Further usefulness of this technique will be apparent when we want to add many sine waves so as to form a "vibration curve." The addition of two sine waves of different amplitudes is illustrated in the problems. Phasors are most often encountered in electrical engineering, where the pertinent sinusoidal functions describe currents and voltages.

Let us return now to the standing wave in its complete form:

$$y(x, t) = 2A \sin\left(\frac{2\pi}{\lambda} x\right) \cos\left(\frac{2\pi}{\lambda} ct\right).$$

We picture this as a stationary sine wave, with amplitude varying (co)sinusoidally with time. (3.4–3.7)

3.5 *Harmonics*

An important kind of standing wave is the one on a string of finite length. On such a string there must be either nodes or antinodes at each end, depending on the boundary conditions. Consider, for instance, a violin string of length L, fixed at both ends. The only waves possible here (any others would quickly die away) are standing waves with nodes at $x = 0$ and $x = L$. The possible wavelengths of such waves belong to an infinite set: $\lambda = 2L/m$, where $m = 1, 2, 3, \ldots$. Of these, the longest is $\lambda = 2L$, with nodes only at the ends. This is called the "fundamental" wave. The next, $\lambda = L$, is called the "second harmonic" (or "first overtone") of the fundamental. The second harmonic corresponds to a frequency double that of the fundamental (an octave, in musical terms).

Systems which support standing waves of this kind are said to be "resonant". Most musical instruments use such resonance to select waves of the desired frequencies. The number and amplitude of the harmonics present contribute to the distinctive nature of the different instruments. Resonant systems also abound in physics. We will use one extensively in Chapter 5. One conceptualization of the energy levels of quantum mechanics involves resonant standing (electron) waves.

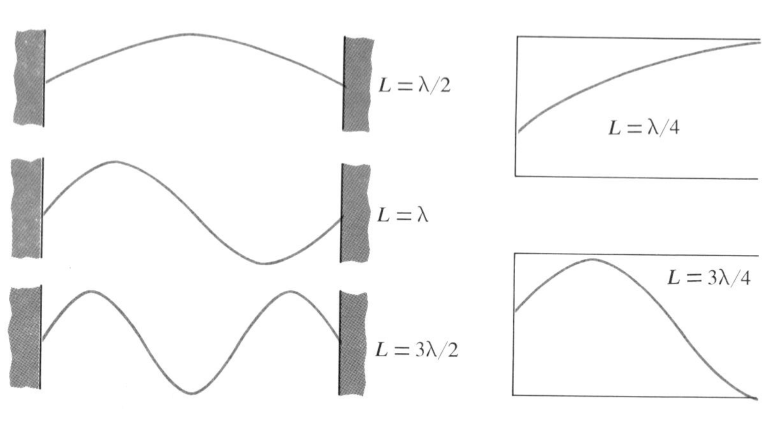

Figure 3.10: (a) Fundamental second, and third harmonics on a string with fixed ends. (b) Fundamental and second harmonic in an organ pipe with one end closed. Pressure forms a node at the closed end and an antinode at the open end.

3.6 *Beats*

A product of two sinusoidal functions, similar to those discussed in Section 3.3, occurs when we add two waves which differ slightly in frequency, but which are traveling in the same direction. Thus

$$A \sin 2\pi\left(\frac{x}{\lambda} - \nu_1 t\right) + A \sin 2\pi\left(\frac{x}{\lambda} - \nu_2 t\right) = 2A \sin 2\pi\left(\frac{x}{\lambda} - \bar{\nu} t\right) \cos 2\pi\left(\frac{\Delta\nu}{2}\right)$$

where

$$\bar{\nu} = \frac{\nu_1 + \nu_2}{2} \qquad \text{and} \qquad \Delta\nu = \nu_1 - \nu_2 .$$

This is the phenomenon of *beats*, which is familiar to musicians. If ν_1 and ν_2 are nearly equal, as they are when we tune two instruments together, then $\bar{\nu}$ is nearly the same as ν_1, or ν_2, while $\Delta\nu/2$ is the frequency with which the tone grows and dies away in amplitude. We say that the term in $\Delta\nu$ *modulates* the main sine wave, the "carrier" wave in communications. Taking the term from music, $\Delta\nu$ is called the *beat frequency*. This is because there are $\Delta\nu$ zeros for every $\Delta\nu/2$ cycles, and it is the zeros that we hear and translate as beat notes. For instance, A above middle C has the frequency 880 Hz, and the note B above A is 988 Hz. Together these give a beat note of 108 loudnesses per second, which sounds like a slightly flat A that is two octaves below middle C. This is the example shown in Figure 3.11. (3.8–3.10)

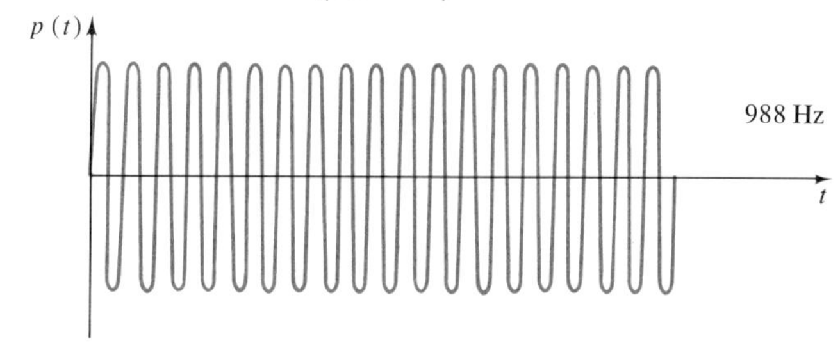

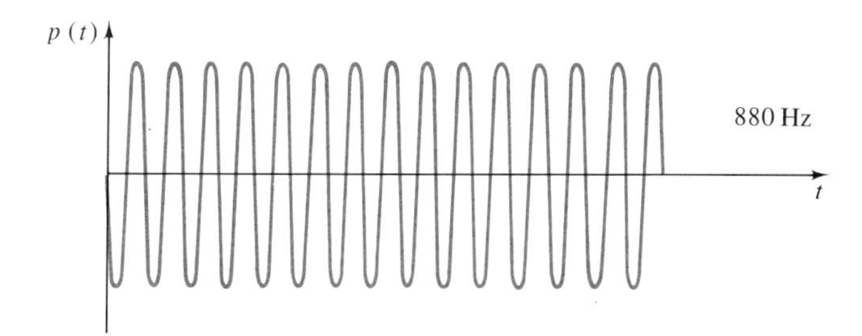

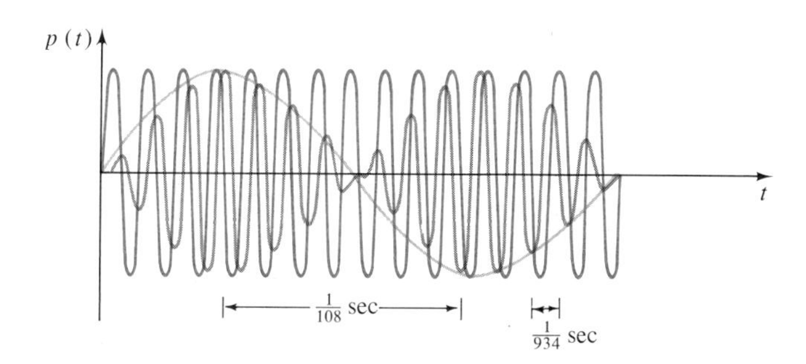

Figure 3.11: Beats.

3.7 *Group velocity*

We have seen that *beats* result when we add two waves of slightly different frequency but of the same wave speed. Now suppose we consider a real material, like glass, for which the wave speed is different for different frequencies: $c = c(\nu) = c(1/\lambda)$. We call such a material *dispersive*, thinking of the behavior of a prism in dispersing the spectrum; we will study this in detail shortly. Let us see what becomes of two waves of different frequency in a dispersive medium:

$$y_1 = A \sin 2\pi\left(\frac{x}{\lambda_1} - \nu_1 t\right), \qquad y_2 = A \sin 2\pi\left(\frac{x}{\lambda_2} - \nu_2 t\right).$$

This is interesting primarily when $\nu_1 \simeq \nu_2$ (hence, $\lambda_1 \simeq \lambda_2$).

$$y_T = 2A \sin\left[\pi\left(\frac{1}{\lambda_1} + \frac{1}{\lambda_2}\right) x - \pi(\nu_1 + \nu_2)t\right]$$
$$\times \cos\left[\pi\left(\frac{1}{\lambda_1} - \frac{1}{\lambda_2}\right) x - \pi(\nu_1 - \nu_2)t\right].$$

As before, this has the functional form of a sine wave with the average frequency and wavelength, modulated by a cosine wave having the difference in frequencies. The sine-wave "carrier" travels with speed

$$v_w = \frac{\text{coefficient of } t}{\text{coefficient of } x} = \frac{\nu_1 + \nu_2}{(1/\lambda_1) + (1/\lambda_2)} \equiv \bar{\nu}\bar{\lambda}.*$$

Figure 3.12: Wave and group velocities for the superposition of two sine waves.

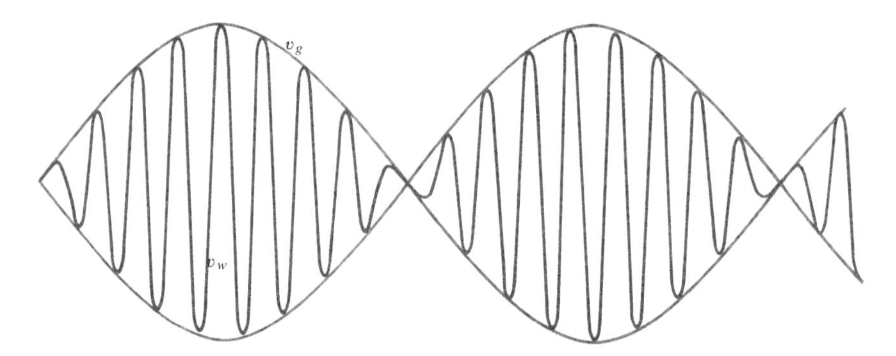

This is the speed with which one must move to keep up with the average-frequency wave. It is called the *wave speed*, and is the speed we have always used so far. But the modulation envelope (the cosine term) also progresses through the medium, and its speed is

$$v_g = \frac{\text{coefficient of } t}{\text{coefficient of } x} = \frac{\nu_1 - \nu_2}{(1/\lambda_1) - (1/\lambda_2)} = \frac{\Delta\nu}{\Delta(1/\lambda)}.$$

* This defines $\bar{\lambda}$:

$$\bar{\lambda} = \frac{1}{2}\frac{\lambda_1 \lambda_2}{\lambda_1 + \lambda_2}.$$

It is really the $1/\lambda$ that gets averaged.

This speed is generally much less than the wave speed. It is called the *group velocity*, and can be written

$$v_g = \frac{\partial \nu}{\partial(1/\lambda)},$$

in the limit as $\Delta\nu \to 0$, and $\Delta(1/\lambda) \to 0$. Pure sine waves are in fact rare in nature, and most media are dispersive; so the group velocity often describes major features of what we see. For instance, throw a rock in a pool and watch the ripples. Here $v_w > v_g$, so the ripple

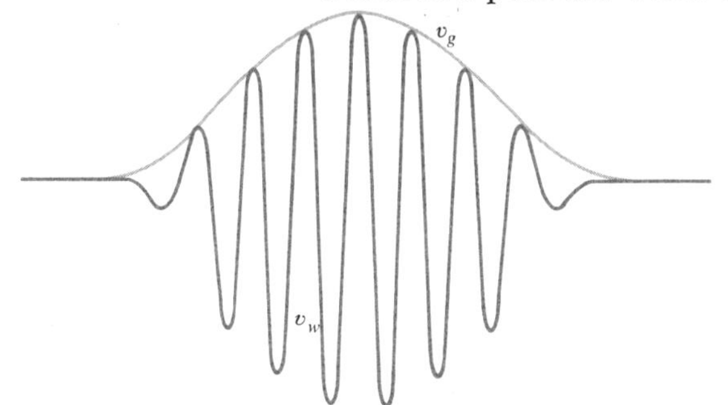

Figure 3.13: Wave and group velocities in a wave packet (ripple).

seems to grow from the rear of the "group," run through it, and vanish ahead. This happens because the ripples are actually a pulse made up of waves whose frequencies cluster around the average frequency, with those closest to the average having the largest amplitudes. Only near the center of the group do the wavelets superpose in such a way that the various disturbances do not cancel. (3.11–3.12)

Such a pulse is one way of thinking of a photon, the particle-like entity that is used to describe light in some situations. The closer the component wavelets bunch in frequency, the longer the group. We can see that this will be the case with the two component

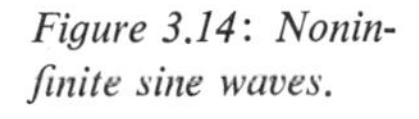

Figure 3.14: Noninfinite sine waves.

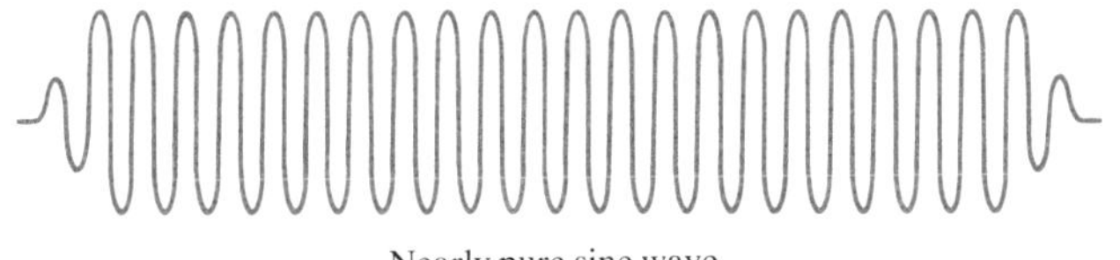

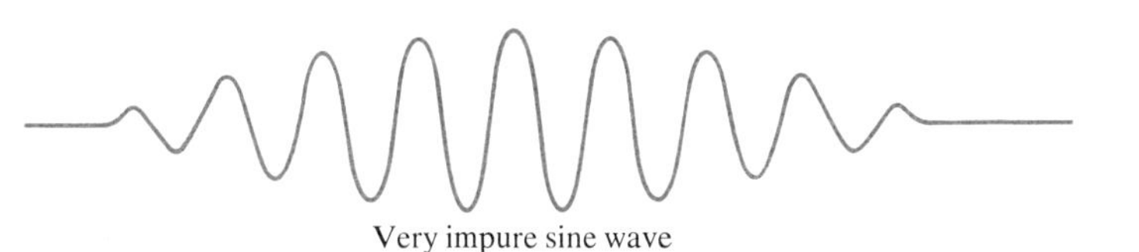

wavelets of our first example, since the smaller $\Delta\nu$ and $\Delta(1/\lambda)$, the longer the "beat". This result is strictly a consequence of the formal mathematics involved in describing the group, and is treated in more detail in Appendix H. However, the same conclusion can be reached for a photon, with a physical argument based on the Uncertainty principle, which states that *a relationship exists between the inherent indefiniteness of the position of a particle and the indefiniteness of its momentum*: $\Delta x\, \Delta p \simeq h$.* For light, this will lead us to $\Delta x\, \Delta\nu \simeq c$. This clearly says the same thing as the mathematics prove: A small spread in frequencies means a large Δx, which we identify as the length of the group. This is why we think of the photons of light emitted by atoms at specific frequencies as about a meter long. When we study interferometry, we will see how to measure this length and demonstrate the relationship. (3.13–3.14)

Appendix H deals with these matters in some detail.

EXERCISES

1. Two sine waves exist on a string:

$$y_1(x, t) = A \sin 2\pi\left(\frac{x}{\lambda} - ct\right),$$

and

$$y_2(x, t) = B \sin 2\pi\left(\frac{x}{\lambda} - ct\right).$$

Show that if $y_T(x, t) = y_1(x, t) + y_2(x, t)$, then $\langle(\text{PE})_T\rangle_{\text{av}} \neq \langle(\text{PE})_1\rangle + \langle(\text{PE})_2\rangle$.

2. Draw a picture like that of Figure 3.1, showing the encounter of a square pulse $y_1(x - ct)$ and a triangular pulse $y_2(x - 2ct)$.

3. Repeat Problem B for $y_1(x - ct)$ and $-y_1(-x - ct)$. Show the point which does not move. Do forces act on it?

4. A guitar string of length L is plucked *near* its center. Waves are reflected from both ends so that only standing waves with nodes at $x = 0$ and $x = L$ persist. What frequencies are present? If the string is plucked *exactly* at the center, what frequencies are present?

* $h = 6.625 \cdot 10^{-34}$ J-sec. This is known as *Planck's constant*.

5. Draw the phasors representing the waves of Problem 1, Chapter 2, at $t = 31$ sec, 32 sec, 33 sec; and of Problem 3, Chapter 2, at $x = 31$ cm, 32 cm, 33 cm.

6. How close in frequency can two radio stations be if we are not to hear their beat note?

PROBLEMS

3.1 Two wave pulses are described by

$$y_1(x, t) = \exp[-(x + ct)^2], \qquad y_2(x, t) = \frac{-1}{1 + (x - ct)^2}.$$

(a) Find y_1, y_2, and $y_1 + y_2$ at $(x, t) = (0, 0)$.

(b) Find the potential-energy density at $(x, t) = (0, 0)$ for each pulse alone and for the two together.

3.2 Two waves are described by

$$f_1(x, t) = 3 \exp\left[-\frac{(x + ct)^2}{25 \text{ cm}^2}\right] \quad \text{and} \quad f_2(x, t) = 2 \sin\left[\frac{2\pi}{4 \text{ cm}}(x - ct)\right],$$

between $x = (ct - 8)$ cm and $x = (ct + 8)$ cm. Elsewhere

$f_2 = (x, t) = 0$ and $c = 6$ cm/sec.

Graph the sum of these functions: $f_T = f_1 + f_2$, at $t = -2$ sec, $t = -\frac{1}{2}$ sec, and $t = 0$.

3.3 For the waves of Problem 1, graph $y_1(0, t) + y_2(0, t)$ and $y_1(1, t) + y_2(1, t)$, where $c = 2$ from $t = -10$ to $t = +10$.

3.4 Two sound waves are described by

$$p_1(z, t) = p_0 \cos\left[\frac{2\pi}{\lambda}(z - ct) + \frac{\pi}{4}\right],$$

$$p_2(z, t) = \frac{1}{2} p_0 \sin\left[\frac{2\pi}{\lambda}(z - ct) + \frac{\pi}{4}\right].$$

(a) What is the total pressure at $z = \lambda$, $t = 0$?

(b) What is the total pressure at $z = \lambda$, $t = \tau/3$?

(c) Use a phasor diagram to describe *qualitatively* the pressure at $z = \lambda$ for all t.

3.5 Two ocean waves are described by

$$h_1(x, t) = h_0 \sin\left[\frac{2\pi}{\lambda}(x + ct) + \frac{5\pi}{4}\right]$$

and

$$h_2(x, t) = \frac{1}{2} h_0 \cos\left[\frac{2\pi}{\lambda}(x + ct) + \frac{5\pi}{4}\right].$$

(a) What is the total height at $x = \lambda/2$, $t = 0$?
(b) What is the total height at $x = \lambda/2$, $t = 2\tau/3$?
(c) Use a phasor diagram to describe *qualitatively* the height at $t = \tau$, for all x.

3.6 Ocean waves are reflected from a smooth vertical sea wall (equivalent to a string with a free end). Choose some reasonable parameters for the waves and describe the motion (velocity, acceleration) and energy of a buoy floating on the surface at various distances offshore.

3.7 Stand with a smooth wall on your right, and hold a tuning fork in your left hand. One ear hears nothing; the other is where it hears the note loudest. What frequencies are possible for the fork?

3.8 A familiar (and useful) system of beats is a *moiré* pattern. Consider two picket fences, one with boards (and spaces) 50 mm wide, and another 51 mm placed next to it. As you drive past at 20 km/hr, how frequently can you *not* see through the two fences?

Top view of fence

3.9 Find the beat frequency between the note A and A-sharp two octaves above middle C. To what note is the beat closest?

3.10 Two identical strings are attached to a wall at $x = 0$. A sinusoidal wave is sent down each.

$$y_1 = A \sin 2\pi\left(\frac{x}{\lambda_1} + \nu_1 t\right), \qquad y_2 = A \sin 2\pi\left(\frac{x}{\lambda_2} + \nu_2 t\right),$$

where $\nu_1 = 0.7\nu_2$. After the waves are reflected from the ends, standing waves are present. Find:

(a) The kinetic energy per unit length at a point $1.75\lambda_1$ from the wall, on string 1.

(b) The kinetic energy per unit length at a point the same distance ($1.75\lambda_1$) from the wall, on string 2.

(c) The distance from the wall at which nodes on the two strings coincide.

3.11 Small ripples over deep water travel at a speed $c = \sqrt{2\pi\sigma/\lambda\rho}$, where σ is the surface tension and ρ is the density of the water.

(a) What is the group velocity for such waves?

(b) If two equal sinusoidal waves of wavelength 10 cm and 11 cm are present, how fast does the node in their beat pattern travel?

3.12 Large waves over deep water travel at a speed $c = \sqrt{g\lambda/2\pi}$, where g is the acceleration due to gravity.

(a) What is the group velocity for such waves?

(b) If two equal sinusoidal waves of wavelength 10 m and 11 m are present, how fast does the node in their beat pattern travel?

3.13 In Appendix H, Figure H.2 is shown a series of wave pulses resulting from the addition of many waves, each differing in frequency from the next by $\delta\nu$.

(a) Identify the wave and group velocities.

(b) Which velocity corresponds to the spread in frequency multiplied by the spread in space?

3.14 In Appendix H, Figure H.5 is shown a wave pulse resulting from the superposition of an infinite number of sine waves, all of infinitesimally different frequencies within the range $\nu_0 - \frac{1}{2}\Delta\nu$ to $\nu_0 + \frac{1}{2}\Delta\nu$.

(a) Identify the wave and group velocities.

(b) Which velocity corresponds to the spread in frequency multiplied by the spread in space (the "size" of the pulse)?

4

Electromagnetic waves, energy and momentum, doppler effect

We are now ready to take up electromagnetic waves, with which we will be primarily concerned from now on. These differ from the simple waves we have studied so far, since the quantity that varies in time and space is a vector field and can exist in vacuum.

4.1 *Electromagnetic waves*

In this chapter we make extensive use of the results of Appendix D, where Maxwell's equations are applied to derive a wave equation governing the electric and magnetic fields associated with a wave in vacuum. The simplest solution of this equation, the plane wave, describes most phenomena in which we are interested. The task of this chapter is to interpret the solution as a physical entity and to see what it tells us about the behavior of these waves.

Simple solutions to the wave equation in Appendix D are

$$\mathbf{E} = \hat{\mathbf{i}} E_0 \sin \frac{2\pi}{\lambda} (z - ct),$$

$$\mathbf{B} = \hat{\mathbf{j}} B_0 \sin \frac{2\pi}{\lambda} (z - ct).$$

Since we start with physical statements, Maxwell's equations, the derivation leaves us with certain physical restrictions on these waves:

1. The **B** wave is *in phase* with the **E** wave.
2. The direction of travel of the waves is $\mathbf{E} \times \mathbf{B}$ (here $\hat{\mathbf{k}}$).
3. $B_0 = E_0/c. \quad c^2 = 1/\mu\epsilon.$

Since one of these waves determines the other, we need only discuss one of them explicitly. We usually choose **E**.

The waves described here are known as *plane waves* because the value of **E** is the same everywhere on a given x, y plane (that is, for a given value of z). For instance, **E** at $(x, y, z) = (6, 5, 10 \text{ m})$ is identical to **E** at $(23.4, -2, 10 \text{ m})$, or any other pair of values for x and y: $(x, y, 10 \text{ m})$.

E changes with time so that all the (identical) values of **E** on one x, y plane change synchronously. The wave progresses in a direction perpendicular to the plane. "Parallel" light is thus necessarily made up of plane waves. The idea of a plane wave presupposes an infinite plane, and we will see that boundaries will affect the wave seriously. For the moment, however, think in terms of infinite waves. (4.1–4.3)

The plane containing the wave crest is called a *wave front.* We could just as well use the plane containing $\mathbf{E} = -\frac{1}{2}E_0\hat{\mathbf{i}}$, or some other value. The crest is convenient to visualize.

Possible frequencies for electromagnetic waves extend from zero to infinity. Some of these are listed in the following table along with the wavelength in vacuum, which is obtained by using the measured value of c in vacuum: $c = 3 \times 10^8$ m/sec.

Frequency, Hz	Wavelength in vacuo	Observed physical effect
0–10^6	∞–few meters	Long-wave radio
10^9–10^{12}	cm, mm	Microwave
10^{14}	micrometers (μm)	Infrared (heat)
10^{15}	$\frac{1}{2}$ μm or 5000 Å	Visible light
10^{16}	nanometers (nm)	Ultraviolet
10^{18}	Å	x rays
$>10^{20}$	<Å	γ rays

The visible region covers roughly the range 0.75 μm to 0.36 μm. These wavelengths are correlated with the colors we perceive, the short wavelengths falling at the violet end of the *spectrum*, and the long ones at the red end.

4.2 *Energy* Electromagnetic waves carry energy. Anyone who has been sunburned can attest to this. The energy per unit volume is calculated as

$$\mathscr{E}=\frac{U}{V}=\frac{\epsilon}{2}E^2+\frac{1}{2\mu}B^2.$$

For the simple plane waves we will study,

$$\mathscr{E}=\frac{\epsilon}{2}E_0{}^2\sin^2\left[\frac{2\pi}{\lambda}(z-ct)\right]+\frac{1}{2\mu c^2}E_0{}^2\sin^2\left[\frac{2\pi}{\lambda}(z-ct)\right]$$

$$=\epsilon E_0{}^2\sin^2\left[\frac{2\pi}{\lambda}(z-ct)\right].$$

Averaging over time or space (see Appendix A),

$$\langle\mathscr{E}\rangle_{\text{av}}=\frac{1}{2}\epsilon E_0{}^2.$$

Note that the **E** and **B** waves each carry half the energy. As with the wave on the string, the average energy content in the electromagnetic wave is proportional to the *square* of the field *amplitude*. This, of course, means that **E** superposes, but that $\mathscr{E}$ does not.

We also want to define a light *intensity* I equal to the average power crossing the unit area in unit time:

$$I=\langle\mathscr{E}\rangle_{\text{av}}\cdot c=\frac{1}{2}cE_0{}^2.$$

4.3 *Momentum* The momentum carried by the electromagnetic wave is in the direction of the wave travel and is equal to the energy divided by c:

$$\frac{\mathbf{p}}{V}=\epsilon[\mathbf{E}\times\mathbf{B}]=\frac{\epsilon}{c}E_0{}^2\sin^2\left[\frac{2\pi}{\lambda}(z-ct)\right]\hat{\mathbf{k}}=\left(\frac{U}{cV}\right)\hat{\mathbf{k}}.$$

There are no everyday examples of this phenomenon, and indeed the momentum of light is so small as to be hard to observe. We will use a physical argument to show that pressure must be exerted by light (which means that it carries momentum) and that the amount of pressure requires the momentum cited. (4.4)

Consider an electromagnetic wave absorbed by a surface in an x, y plane. Later we will extend this to reflecting surfaces. How does the absorption take place? Electromagnetic fields interact with charges, so we will concentrate on one such in the surface. When the charge q is subjected to the electric field of the wave, it experiences a force $\mathbf{F} = q\mathbf{E} = m\mathbf{a}$ and is accelerated (from rest, for simplicity). So we can write

$$\mathbf{a} = \left(\frac{q}{m}\right) \mathbf{E}_0 \sin\left[\frac{2\pi}{\lambda}(z_0 - ct)\right] \qquad \text{at } z = z_0,$$

the position of the charge. We can abbreviate this as

$$\mathbf{a} = \left(\frac{q}{m}\right) \mathbf{E}_0 \sin \theta(t).$$

If there were absolutely no restraining forces on the charge, it would then acquire the velocity

$$\mathbf{v} = -\frac{q\mathbf{E}_0}{m}\frac{\lambda}{2\pi c} \cos \theta(t).$$

But such forces are never wholly absent, so the charge is held back a little and the velocity lags by a small amount, $\delta\phi$:

$$\mathbf{v} = -\frac{q\mathbf{E}_0}{2\pi\nu m} \cos[\theta(t) + \delta\phi].$$

Now the electric field is doing work here (indeed, that is how the wave gets absorbed), and we can calculate the work done in 1 cycle:

$$W = \int \mathbf{F} \cdot d\mathbf{l} = \int_{1\text{ cycle}} \mathbf{F} \cdot \mathbf{v}\, dt.$$

Then the average power absorbed (the work done per unit time) is

$$\langle P \rangle_{\text{av}} = \frac{1}{\tau}\int_0^\tau \mathbf{F} \cdot \mathbf{v}\, dt = -\frac{1}{\tau}\frac{(qE_0)^2}{2\pi m\nu}\int_0^\tau \sin[\theta(t)] \cos[\theta(t) + \delta\phi]\, dt.$$

Note that if $\delta\phi = 0$, no power would be absorbed. (See Appendix A.)

The magnetic wave also interacts with the charge, exerting a force $\mathbf{F}_m = q\mathbf{v} \times \mathbf{B}$ on it. Again follow the charge for 1 cycle and find

$$\langle \mathbf{F}_m \rangle_{\text{av}} = \frac{q}{\tau}\int_0^\tau \mathbf{v} \times \mathbf{B}\, dt = -\frac{1}{\tau}\hat{\mathbf{k}}\frac{(qE_0)^2}{2\pi\nu mc}\int_0^\tau \sin\theta \cos(\theta + \delta\phi)\, dt.$$

So

$$\langle P \rangle_{\text{av}} = c\langle F_m \rangle_{\text{av}}.$$

(Incidentally, notice that $\langle F_e \rangle_{\text{av}} = 0$.)

Now we have a relationship established between the average

power absorbed and the average force exerted. This relationship is often useful as it stands, but we can state it in terms of more fundamental quantities if we note that

$$P = \frac{dU}{dt} \qquad (U = \mathscr{E} V) \qquad \text{and} \qquad F = \frac{dp}{dt}.$$

Then

$$\langle U \rangle_{\text{av}} = c \langle p \rangle_{\text{av}} \qquad \text{or} \qquad U = cp,$$

since the averaging can be left out in a more rigorous treatment. (4.5)

Now, what about the reflection? If the charge is the "free" charge in a metal, $\delta\phi$ is very small and is due primarily to loss of energy as the charge (being accelerated) emits an electromagnetic wave. Accelerated charge is always the source of electromagnetic waves, a fact which follows from Maxwell's equations. We can accept this here as provable by taking a movie of the absorption process just described and showing the film backwards. All physical laws with which we are familiar are time-reversible. That is, there is no statement that the sequence of events can proceed only in one direction in time. Thus, a sequence which is possible with the film running forward is also possible when it runs in reverse. If the external forces responsible for $\delta\phi$ are sufficiently small, the charge re-emits the energy it is absorbing as a "scattered" wave. In fact, for a completely free charge, this re-emission is the only source of the $\delta\phi$. When many charges are present, the scattering is preferentially backward, and the wave is said to be reflected.

The force due to the emission is in the same direction as that due to the absorption. The reason for this is also seen from the reversed movie: Velocities change their signs when the sign of time is reversed. Thus, $\mathbf{v}$ in our equations changes sign and so does $\mathbf{B}$, whose source currents change sign (the moving charges that make up the currents have velocity). On the other hand, $\mathbf{E}$ has stationary charges for its sources, so its sign is not changed. Thus, the direction of travel of the wave, $\mathbf{E} \times \mathbf{B}$, is reversed, but the direction of the magnetic force, $\mathbf{v} \times \mathbf{B}$ is not. So P changes sign (energy is emitted instead of absorbed), but $\mathbf{F}_m$ does not. The result is that the reflected wave has given up twice as much momentum to the charge as the absorbed wave did. The momentum transfer results in a pressure being exerted on charges, and through them on material bodies; this is known as the *radiation pressure*. (4.6)

4.4 *Photons* When light exerts pressure, the light waves behave exactly like elastic bodies ("billiard balls"). That is, a billiard ball transfers all its momentum (and energy) to a body it hits and sticks to, and transfers twice its momentum to a body that reflects it. Such a comparison suggests a view of the light as composed of particles. This idea is not only fruitful, it seems to be an equivalently good "model" of the observed phenomenon. We are primarily concerned here with the wave picture of light, but let us digress briefly to mention *photons*.

A photon is a "particle" of light. Its energy is $U = h\nu$, where ν is the frequency of the light when we think of it as a wave, and h is a universal constant (Planck's constant): $h = 6.625 \times 10^{-34}$ J-sec. The photon's momentum is then $p = h\nu/c = h/\lambda$. The photon is never observed at rest, and would in fact have zero mass if it were at rest. But since we know its velocity, c, and momentum, we might assign it a mass: "m" $= p/c = h\nu/c^2$. Note that this is suggestive of the relativistic result $U = mc^2$. The first two rules of geometrical optics are consistent with the proposition that light is composed of particles.

The pressure of light leads to some rather interesting calculations. Consider, for instance, a flashlight adrift in space: If the light emits a beam in one direction, with power P for T seconds, its subsequent motion is described by $MV = PT/c$, where M is the mass of the flashlight. All we have done here is to equate the total momentum before the light was emitted (zero) to that afterward ($MV - PT/c$). The details of the process are not important; all we require is that we know the total energy in the final beam, its direction of flow, and how long it lasts. How it was made unidirectional and what frequencies are present are irrelevant details. (4.7–4.9)

For a static situation, consider a piece of metal foil floating on sunlight: The forces to be balanced are Mg, where M is the mass of the foil and the radiation force is $2P/c$. For sunlight, P is 0.135 W for

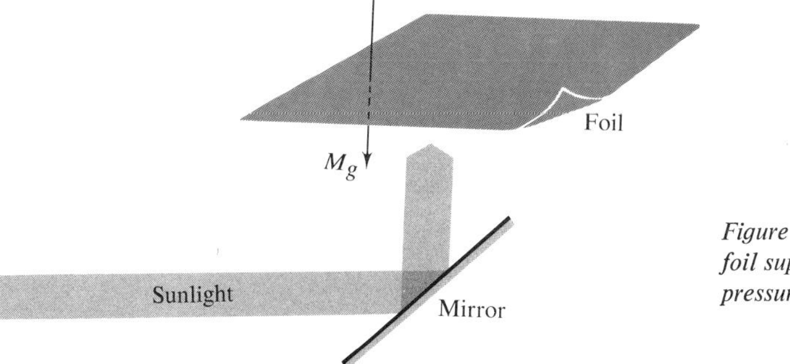

Figure 4.1: Aluminum foil supported by light pressure.

a 1-cm^2 foil. This requires that our foil have a mass (for 1 cm^2) of about 10^{-8} kg. For aluminum, this means a foil about 1000 atoms thick—just feasible, but at the mercy of a myriad of other forces. In the Bibliography, a more realistic device is described for demonstrating light pressure, but the calculation is a bit more involved.

4.5 *Doppler effect*

We have discussed the emission of electromagnetic waves and the fact that reflectors recoil. Let us consider for a moment what we will observe if *any* wave source is in motion, as well as the similar problem of what we observe when we move with respect to the source. We will find that if we are moving relatively toward the source, the measured frequency increases. This is the Doppler effect. One observes it typically in the rising pitch of the horn of an approaching car.

In the case of sound (and, in fact, in all but electromagnetic waves), there is a different result, depending on whether it is the source of the wave or the observer that is moving. For light, this will not be the case. In many situations these distinctions make no difference anyway, since we can use the approximation techniques of Appendix A to show that

$$\frac{\Delta\nu}{\nu} \simeq \frac{v}{c}$$

where $\Delta\nu$ is the change in frequency of the wave, v is the relative

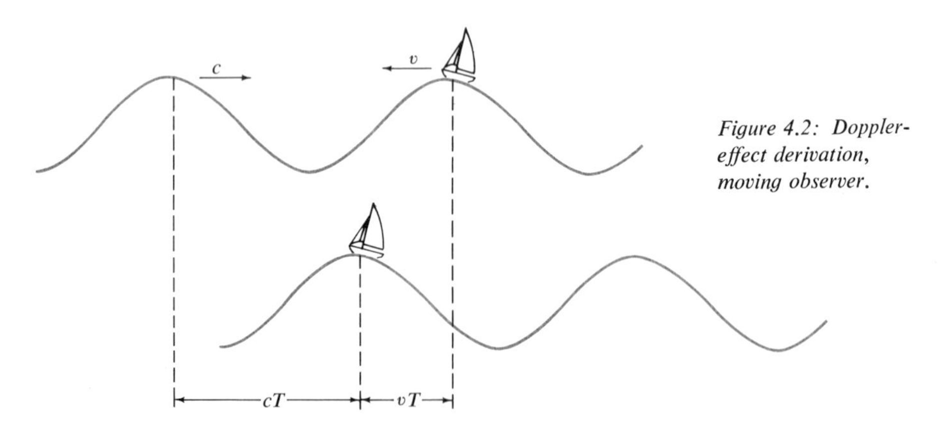

Figure 4.2: Doppler-effect derivation, moving observer.

velocity of source and observer, and c is the wave speed. This approximation holds if $v \ll c$, which is usually the case for light and often for other waves. We will work first with simple material waves.

Consider a boat on the ocean. An observer in the stationary boat finds that $\nu = c/\lambda$. If he now sails toward the source of the waves, he hits the crests more often than when he was sitting still. Suppose the boat moves at speed v relative to the water, and the wave speed is c. A stationary boat goes from one wave crest to the next in a time τ, but the moving boat takes only T for this trip. During this time, the boat has traveled a distance vT relative to the water, and the wave a distance cT. Since the boat has moved a distance λ relative to the wave, we can see that $\lambda = vT + cT$.

If we call ν' the measured frequency with which the boat encounters crests,

$$\nu' = \frac{1}{T} = \frac{v+c}{\lambda} = \nu\left(1 + \frac{v}{c}\right).$$

We need not worry about the sign of the v/c term because common sense should always supply it: If the observer moves into the waves, the frequency goes up.

But suppose it is the source that is moving? Consider, for instance, a moving loudspeaker: It sends out a wave crest and then moves a distance $v\tau$ before sending out the next. So the distance between crests is $c\tau - v\tau = \lambda'$. An observer measures $\nu' = c/\lambda' = \nu/[1 - (v/c)]$. A general expression embodying both results is

$$\nu' = \nu \frac{1 + v_{ob}/c}{1 - v_{sc}/c}.$$

Here the sign of the pertinent v is always chosen so that the frequency increases when source and observer are approaching each other. To remember which v goes in the denominator and which in the numerator, set $v_{ob} = c$. An observer moving away at speed c hears no sound. Similarly, a source moving toward the observer at speed c bunches all its crests together in a sonic boom, which shows up as an infinity in the frequency equation. The wrong choice gives mere factors of 2. (4.10)

Let us now ask the question: What is the speed c relative to? This is the speed of the wave relative to the medium (in the case of sound, air). Can the measured frequency tell us whether the air is in motion? It can, but only if source and observer are in relative motion also. Let the wind speed by v_w. Then the velocities of the wave, the observer, and the source relative to the air are

c, $v_{ob} - v_w$, and $v_{sc} - v_w$, respectively. The measured frequency is

$$\nu' = \nu \frac{c - (v_{ob} + v_w)}{c - (v_{sc} + v_w)},$$

with all velocities positive in the direction of wave travel. If $v_{ob} = v_{sc}$ (as, for instance, when both are zero), $\nu' = \nu$. This merely confirms our common experience that sounds do not change pitch when the wind blows. Of course observer and source may be in relative motion, and then v_w is measurable.

After this introduction, we should now apply the theory to light. But it turns out that light behaves differently. The qualitative features are the same—the frequency increases if source and observer move toward each other—but here there is nothing to distinguish which is in motion. This is a consequence of the fact that no matter what our state of motion, we always find the same value for the speed of light in vacuum. That is, if we measure the speed of the light from some star, both on earth and in a space ship going toward the star at half the speed of light, both measurements give 3×10^8 m/sec. The Doppler equation for light must be derived relativistically, so we give it here without proof: (4.11–4.12)

$$\nu' = \nu \sqrt{\frac{1 + (v/c)}{1 - (v/c)}}.$$

Many phenomena in physics demonstrate the Doppler effect on light. Stationary atoms emit light at specific frequencies, but atoms moving at high speed in a hot gas emit light over a range of frequencies. This is because some atoms are moving toward the observer and some away from him. Thus the narrow frequency range of the stationary atom is said to be *Doppler broadened*. Similarly, the wavelengths from the approaching edge of a rotating star are shifted toward the blue and those from the receding edge toward the red, thus enabling astronomers to measure stellar rotation rates.

The most spectacular optical Doppler effect is the intergalactic red shift. All the galaxies in the universe are flying apart, causing their light to appear redder. The more distant they are, the larger the red shift (that is, the faster they are receding). One consequence of this may be that the nighttime sky is dark. An infinite universe full of emitting stars would send us infinite light. Interposed absorbers would merely heat up and start to emit also. But since the majority of an infinite number of stars have their light infinitely reddened, it makes no contribution to the energy we receive (remember that $U = h\nu$).

EXERCISES

1. If $\mathbf{E}(z_0, t_0) = (16 \text{ V/m}) \sin(0.8)\hat{\imath}$, find $\mathbf{B}(z_0, t_0)$.

2. If $\mathbf{E}(x, y, z, t) = \hat{\jmath}(18 \text{ V/m}) \sin 2\pi(z/7 \text{ m} - t/19 \text{ sec} + 1/8)$, find $\mathbf{B}$(6 m, 8 m, 7 m, 19 sec).

3. A laser emits 1 W of power at wavelength $\lambda = 0.62$ μm in a parallel beam 1 cm^2 in cross section. Find $\mathbf{E}$ in the beam.

4. How much momentum is there in the beam of Exercise 3?

5. How many photons does a 50-kW radio station ($\nu = 10^6$ Hz) emit per second?

6. Show that $\Delta\nu/\nu \simeq v/c$ whenever $v \ll c$, for all possible situations.

PROBLEMS

4.1 If an electromagnetic wave has its electric field vector described by $\mathbf{E}(z, t) = \hat{\jmath}E_c \cos[(2\pi/\lambda)(z + ct)]$, what is $\mathbf{B}$(10 cm, 10 sec)? Take $\lambda = 3.14$ cm.

4.2 A sheet of metal reflects an electromagnetic wave which is incident on it normally (that is, the wave fronts are in the same plane as the metal). If the metal is a perfect conductor, we know that $\mathbf{E}_\text{T} = 0$ in it. This means that the standing wave formed from the incident and reflected waves has an $\mathbf{E}$ field *node* at the sheet. Write the equation for $\mathbf{B}_\text{T}(z, t)$.

4.3 A sheet of metal, like that of Problem 4.4, is in the $z = 0$ plane. *Two* electromagnetic plane waves fall on it normally, with frequencies in ratio $\nu_1/\nu_2 = 13$.

(a) Find the planes for which E nodes coincide; that is, those where $\mathbf{E}_\text{T} = 0$ at all times.

(b) Find $\mathbf{B}_\text{T}(t)$ in these planes.

4.4 We are given the following information about a sinusoidal plane electromagnetic wave:

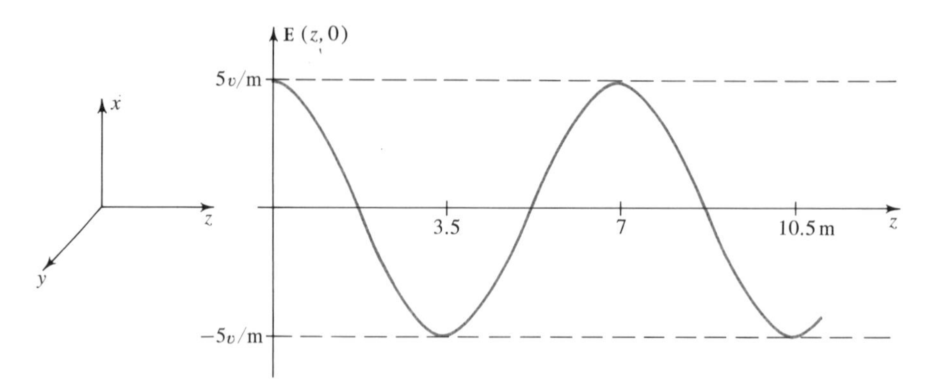

(a) Find $\mathbf{B}(z, t)$.
(b) Find the momentum density in the wave.
(Remember that both **B** and $\mathbf{p}/V$ are vectors.)

4.5 As a sinusoidal plane electromagnetic wave passes a magnetometer, the following chart is made:

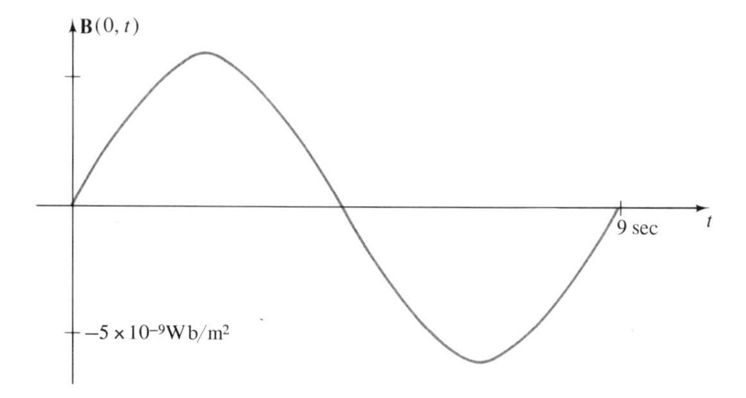

B is along the x axis and the magnetometer can be moved anywhere in the x, y plane without changing its readings.

(a) Find $\mathscr{E}(z, t)$.

(b) Find the force on an absorber in the x, y plane. Remember that **F** is a vector.

4.6 A reflecting particle is attracted to the sun by gravity and repelled by light pressure. The particle is just the right size for the two forces to balance out. Take its specific gravity as 2 and find its diameter. (The result is the same whether the particle is a sphere or a disk. Prove this.) What effect does its distance from the sun have?

4.7 A flashlight of mass M floats in space, at rest with respect to the "fixed stars". It is turned on for 1 hr, emitting 1 W of light in a parallel beam. How fast is it going at the end of the hour?

4.8 A photon of visible light is specularly reflected by an electron. What is the recoil velocity of the electron? What is the energy of the reflected photon? Is the reflected photon still visible?

4.9 Solve Problem 4.8 with an x-ray photon and a proton.

4.10 Atoms of a star in the plane of the ecliptic emit light of characteristic wavelength 0.6 μm (6×10^{-6} m). Measurements of this wavelength are made at the South Pole, and vary in a regular fashion throughout the year. Plot $\lambda'(t)$ over one year.

4.11 The Tsunami river flows south at 10 km/hr. A south wind of 30 km/hr raises waves on the river surface, which travel at 20 km/hr relative to the water. The waves are observed by a skin diver in the river, a balloonist above it, and a fisherman on the bank. What is the ratio of the frequencies the three men observe?

4.12 A laser emits a monochromatic light beam, of wave length λ, which falls normally on a mirror moving at velocity V. What is the beat frequency between the incident and reflected light?

4.13 A train 200 m long is traveling at 72 km/hr. The engineer sees a pig on the track, blows his whistle, and causes the pig to squeal and the farmer to run toward the pig at 5.4 km/hr. The engineer hears the whistle as 100 Hz, and the squeal as 2 kHz. Find the frequencies heard by (a) a brakeman in the caboose, (b) the pig, (c) the farmer.

5
Scattering: index of refraction

In this chapter we investigate the phenomenon of *scattering*, whereby the charges in matter absorb and re-emit electromagnetic waves. From this we hope to understand the microscopic origin of the index of refraction.

5.1 *Scattering*

The term *scattering* is applied generally to situations in which waves encounter an obstacle much smaller than a wavelength, with the result that energy is sent in a new direction. An isolated scatterer

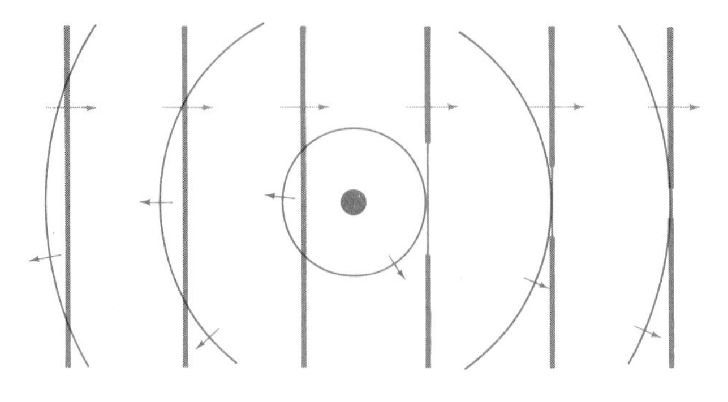

Figure 5.1: Plane wave incident on a single scatterer.

absorbs some energy from an incident plane wave and re-emits it as a spherical wave, possibly with some loss of energy to frictional forces.

For electromagnetic waves, the scatterer must be an electric charge, either free or part of an atom, usually in matter. If we think of material like glass as having many scattering charges, then their combined effect will be the phenomenon of refraction. Before investigating the microscopic aspects of this, let us use the idea of scattering to relate the index of refraction, n, to a change of wave velocity in the medium.

5.2 *Refraction*

In the plane wave of Figure 5.2, we can observe the scatterers in any wave front. In part (a), the medium is homogeneous and the envelope of the scattered "wavelets" (that is, the tangent to them) is the new wave front. (We will discuss below why we do not observe any parts of the wavelets except those going forward.) As long as all scatterers are in the same medium, the wave progresses in a straight

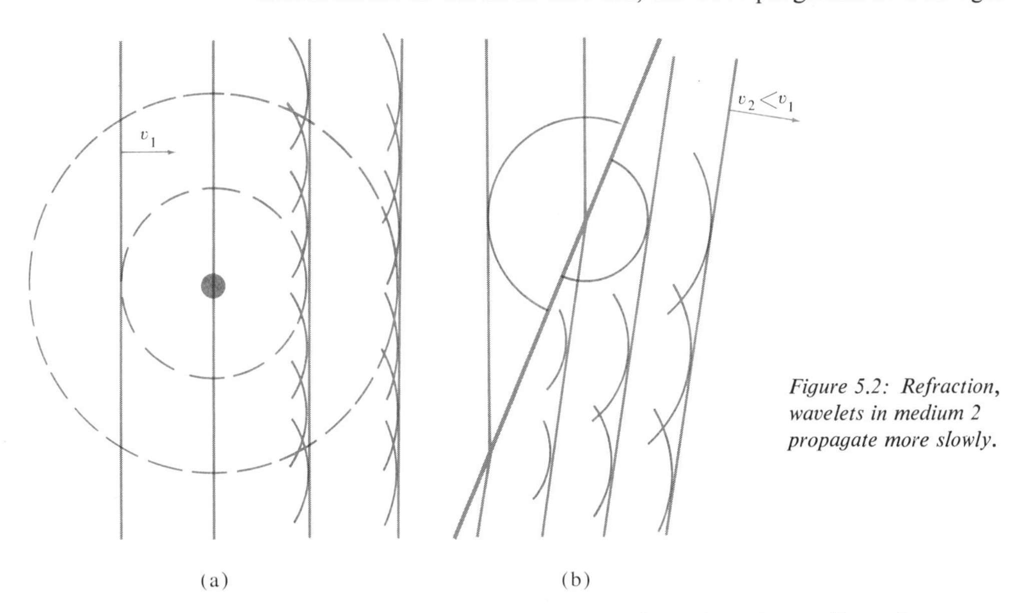

Figure 5.2: Refraction, wavelets in medium 2 propagate more slowly.

line. Figure 5.2(b) shows the situation when the medium changes to one in which the wave speed is different. In this case the wavelets grow at a different rate so that the tangent wave front lies in a new plane and refraction has occurred. Notice that it is the *wave* velocity rather than the group velocity which is measured by refraction. We will explore the microscopic features of the scattering, then, to learn why the velocity changes and to understand the general features of the index of refraction.

5.3 *Index of refraction*

A major effect that we wish to understand is the frequency dependence (dispersion) of the index of refraction. As the light wave propagates through the transparent material, it does so at a speed different from its speed in vacuum (it usually goes slower), and this speed generally varies with the wavelength of the light. We will interpret these effects as the result of the wave's interaction with charges bound in the matter by harmonic forces, forces such as the restoring force in an elastic spring: $F = kx$. For the details of this interaction we draw on Appendix E (Resonance). Our concern here is with the light rather than with the mechanical system.

We will adopt this picture of matter: The charges in the material with which the light wave interacts are bound to fixed points in such a way that they have one or more characteristic resonant frequencies. Each resonance is important over a range of frequencies determined by the frictional forces. The charges can absorb and then re-emit the light (that is, scatter it) in all directions. If the light has a frequency near one of the resonant frequencies, some of its energy is permanently absorbed, going into the frictional forces. The wave is then scattered with some loss of amplitude. If the light frequency is not close to a resonance, the matter is transparent and nearly all the energy is re-emitted, possibly with some phase shift.

The phase shift is the crucial question, and we will consider it presently. However, let us consider briefly why, in materials like glass, the light appears to be scattered only forward. The charges we are dealing with are spread evenly through space, a few angstroms apart (one or more for every atom). Thus they are spaced much less than a wavelength of visible light, homogeneously as far as the light is concerned. In this case the total wave, E_T, scattered backwards or sideways is zero. The homogeneity is spoiled at the glass-air interface, and we do indeed see such surfaces by scattered light. Inhomogeneities in the matter, such as dust particles in water, flaws in glass, and density fluctuations in the upper atmosphere, also scatter enough light sideways to become visible (this last causes the bright daytime sky).

The homogeneous regions, however, scatter light only forward, and this constitutes one of the phenomena which we expect to explain by a consideration of the phase shifts. The argument can be understood from Figure 5.3. Within the short distance Δz, we might see two possible scatterers, A and B, one a bit farther in than the other. The phasor diagrams tell the story.

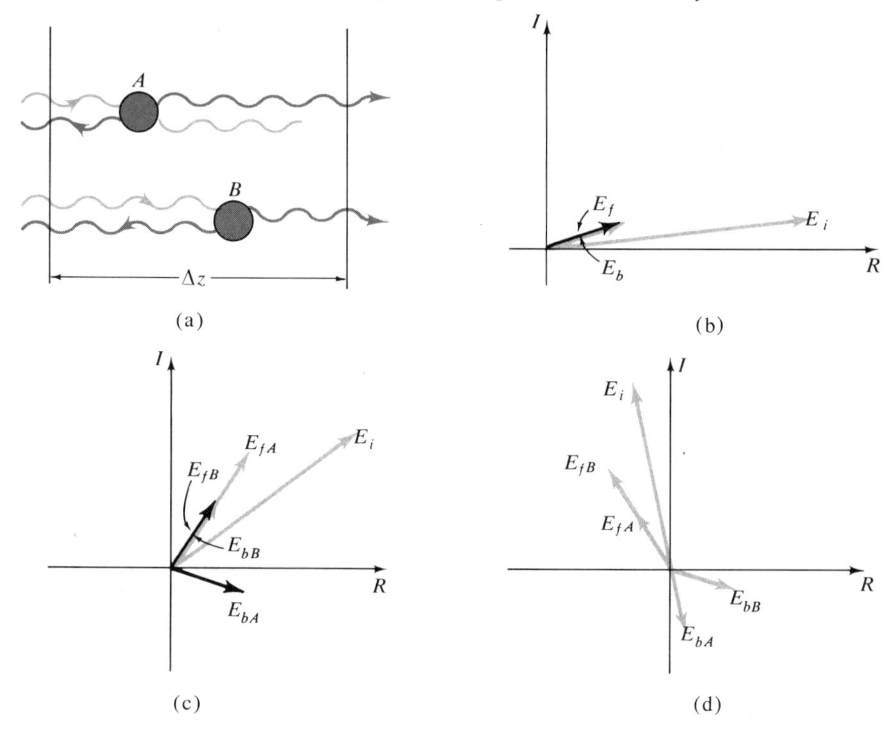

Figure 5.3: Randomly situated scatterers scatter only forward, as a group.

In Figure 5.3(b), we see the phasors at the position of charge A: We show phasors for the incident wave and the waves scattered forward and back from A. Figure 5.3(c) shows these phasors at the position of the second scatterer: The forward wave preserves its phase relationship to the incident wave and to the second forward-scattered wavelet. The backward one, however, rotates the other way so that it is not in phase with the new back-scattered wave. Wave E_{bA} does not actually exist at the position of the second scatterer z_B, but we extrapolate it back to compare phases. Figure 5.3(d) shows the situation at the position $z = \Delta z$. The two forward waves are in phase, and therefore add up. The two back-scattered waves are out of phase by an amount depending on the distance between atoms A and B, a distance which is different for every pair of scatterers. Thus, the forward waves from N randomly distributed scatterers are all in phase, and all add up. But the backward waves have phases distributed randomly so that their resultant is zero, as shown in Figure 5.4.

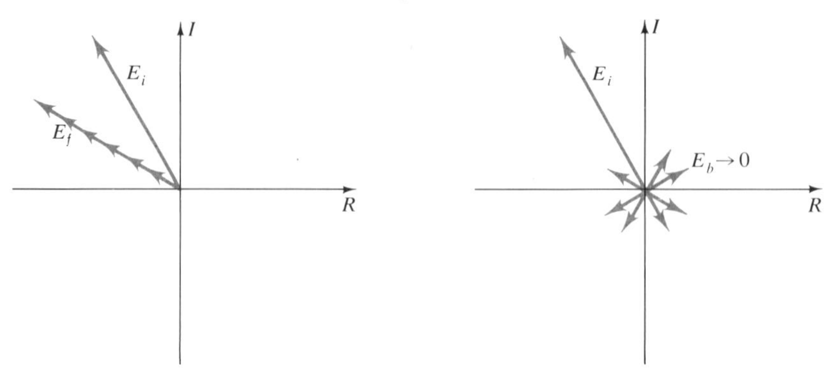

Figure 5.4: Forward wavelets from random scatters all have the same phase. Backward wavelets have random phases.

The result of all this is that the only detectable scattered wave is in the forward direction, and no energy goes backward. The same argument can be used to make the case against the side-scattered wave. This, incidentally, is a fine example of the use of phasors. They enable us to extract a quantitative result from a qualitative argument.

The argument just given can also show us why the wave velocity changes in the medium: The only net effect that the scattering has on the wave is to advance or retard its phase by $N\,\Delta\phi$, where N is the number of times it is scattered (say, in unit length). Thus, the resultant wave, having traversed unit length of the material, emerges with some net phase change, which we could calculate from N and $\Delta\phi$. Suppose $N\,\Delta\phi$ represents a lag in phase; we could merely say that the wave crest appeared later than it would have appeared if the wave had traveled in vacuum. But this is precisely what would result if the wave moved more slowly through the medium! And so we have our index of refraction. Explicitly, the wave encounters bound charges and is re-emitted by them with the form

$$E_{\text{scat}} = 2\pi\nu K \cdot E_0\left[\chi' \cos 2\pi\left(\frac{x}{\lambda} - \nu t\right) - \chi'' \sin 2\pi\left(\frac{x}{\lambda} - \nu t\right)\right],$$

where the χ' and χ'' are found in Appendix E, and K is a constant. The phase of the wave scattered from the sheet of bound charges is the same as the *velocity* of the charges; that is, it is 90 degrees out of phase with the wave scattered from a single charge.

Now, if we think of the medium as having an index of refraction $n = c/v_w$, we can find another expression for the scattered wave. After passing through a thickness L of the material, the original wave has become

$$E(L,t) = E_0 \sin 2\pi\nu\left[\frac{L}{v_w} - t\right]$$

$$= E_0 \sin 2\pi\nu\left[\frac{L}{c} - t + \left(\frac{L}{v_w} - \frac{L}{c}\right)\right]$$

$$= E_0 \sin 2\pi\nu\left[\frac{L}{c} - t + \frac{L}{c}(n-1)\right].$$

Just as we expected, the slowdown of the wave, due to the index of refraction, is equivalent to a phase shift. We can also write this as

$$E(L,t) = E_0 \sin\left[2\pi\nu\left(\frac{L}{c} - t\right)\right] \cos\left[2\pi\nu\frac{L}{c}(n-1)\right]$$
$$+ E_0 \cos\left[2\pi\nu\left(\frac{L}{c} - t\right)\right] \sin\left[2\pi\nu\frac{L}{c}(n-1)\right].$$

Our E_{scat} has been scattered just once, so we must make L very small so that $E(L, t)$ will have encountered only a single sheet of charges. If there are N charges along a line of unit length, then L must equal $1/N$, and it *is* very small.

Now we equate E_{scat} and $E(L, t)$ to see that

$$2\pi\nu K E_0 \chi' = E_0\left[\sin 2\pi\nu\frac{L}{c}(n-1)\right].$$

Since L is very small, $n = 1 + K'\chi'$, where K' involves various constants pertinent to the specific system:

$$K' = K\frac{c}{L} \propto N.$$

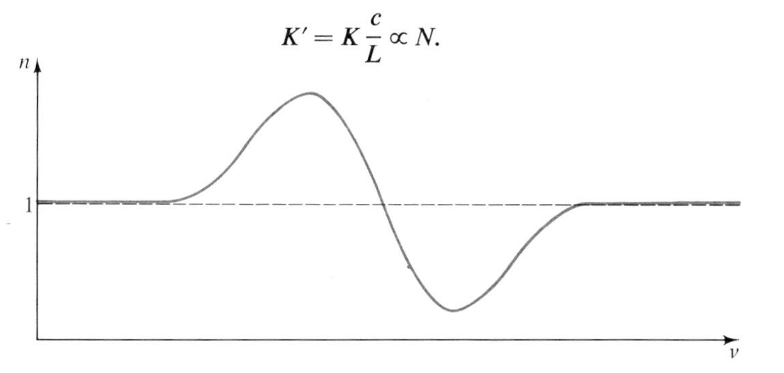

Figure 5.5: The index of refraction as a function of frequency.

The index of refraction follows a curve like that of Figure E-3 (Appendix E), where we have anticipated by calling one of the curves the *dispersion*. Accompanying the dispersion, we may expect to find an absorption peak, as discussed in Appendix E. This is seen to be the case, as illustrated by the curves for glass and fuchsine dye shown in Figure 5.6. In Appendix F the same result is shown to come from simple electromagnetic theory, starting from the same resonant scatterer. As we expected, the behavior of a single bound charge is reflected in the index of the material. The kind of calculation we have just done is similar to techniques used in discussing the scattering of subatomic particles in modern physics. As we often see, the waves of quantum mechanics behave like other waves and are treated similarly.

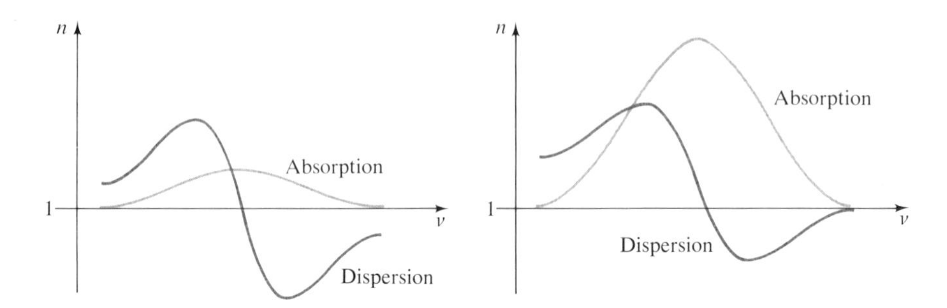

Figure 5.6: (a) Dispersion and absorption in glass. (b) Dispersion and absorption in fuchsine dye.

5.4 *Birefringence*

Until now we have regarded the choice of the direction of **E** as arbitrary. But there are materials in which we measure a different value of the index of refraction, depending on which way **E** points relative to certain directions in the solid. Such a material is said to be *birefringent*, or doubly refracting. We will use birefringent materials extensively in our investigation of polarized light. The cause of the phenomenon is fairly easy to find: Suppose that a charge is bound more tightly to its neighbors in the x direction than in the y direction: Again, we represent the restoring forces by springs, but now $K_x \neq K_y$. The kind of thing which can cause such an asymmetry might be that the neighboring ions are closer in one direction than the other. Now different spring constants, K, lead to different resonant frequencies so that the charge will be in resonance at one frequency if **E** is in the

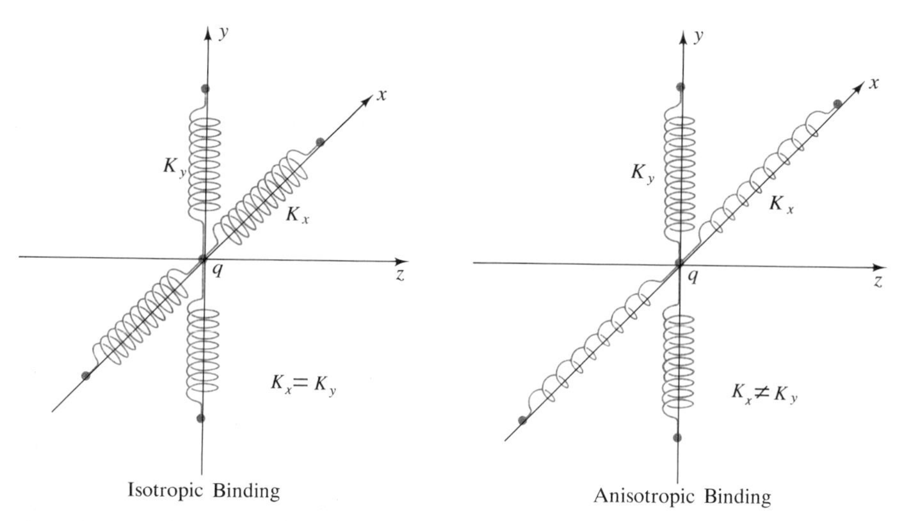

Figure 5.7: Birefringent scatterer.

x direction and at another frequency if it is in the y direction. Note that in either case the wave propagates in the z direction. The two resonant frequencies lead in turn to two indices, as shown in Figure 5.8.

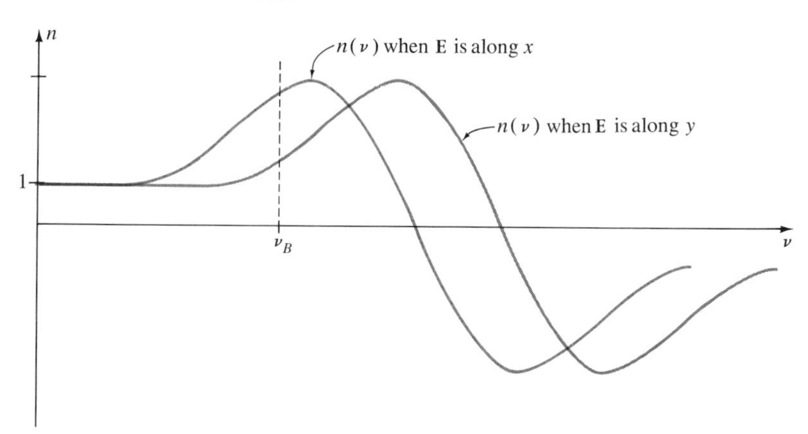

Figure 5.8: A material that is birefringent at the frequency ν_B.

Suppose the frequency of our light is ν_B. Then the measured index at that frequency is much greater if **E** is along x than if **E** is along y, in the example here. Since our labeling of the axes x and y is quite arbitrary, it might be better to call them the *slow* and *fast* axes and say $n_{\text{slow}} > n_{\text{fast}}$. Notice that it is *not* the direction of travel that determines which index applies; here the wave travels in the z

direction for both. Rather, it is the direction along which **E** lies, called the direction of *polarization* of the wave. Thus, if **E** is along the slow axis, the wave travels through the medium with the speed $v = c/n_{\text{slow}}$, the slower of the two speeds. Of course **E** may lie along some intermediate axis, in which case we divide it into a slow and a fast component, which get out of phase. This is treated in the next chapter.

5.5 Dichroism

A less common variant of the two indices of a birefringent material occurs when the light frequency falls near the frequency marked ν_D, as in Figure 5.9. In this case, the material is strongly absorbing for one polarization and transparent for the other. Such a material is said to be *dichroic*, and is familiarly represented by the commercially important sheet polarizer. Such a material appears opaque to light having its **E** vector in one direction, and is transparent if **E** is in the other direction.

Figure 5.9: A material that is dichroic at the frequency ν_D.

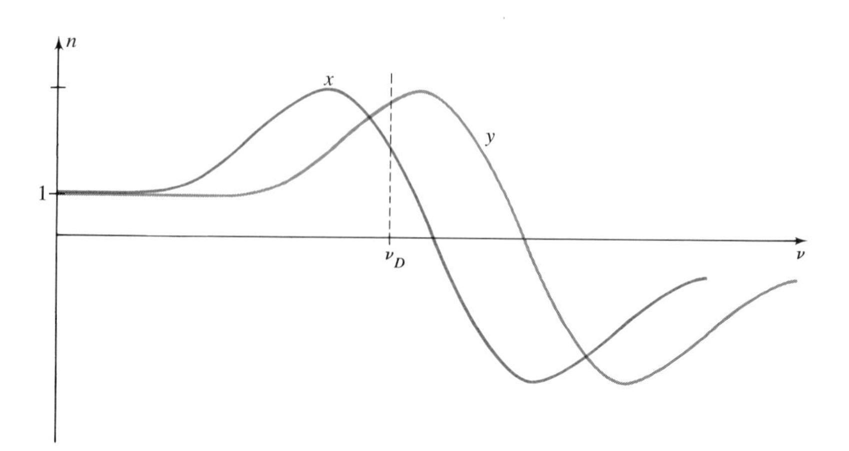

EXERCISES

1. Show that the construction of Figure 5.2 results in the law of refraction if we let $n = c/v$.

2. A crystal scatters preferentially in other directions besides forward. Draw a square "lattice" of scatterers and find two such directions.

3. A hydrogen atom consists of an electron bound to a (relatively) fixed proton. If the electron moves from its equilibrium position ($r_0 = 0.5$ Å), the restoring force is

$$\Delta F = F - F_0 = \frac{e^2}{4\pi\epsilon_0(r_0+\delta)^2} - \frac{e^2}{4\pi\epsilon_0 r_0{}^2} \simeq \frac{2e^2\delta}{4\pi\epsilon_0 r_0{}^2}.$$

Find the resonant frequency.

4. Use handbook values to plot $n(\lambda)$ for glass from $\lambda = 10$ μm to $\lambda = 0.1$ μm.

5. Find the difference in energy density for electromagnetic waves in water and air. Express this as a change in ϵ and find the dielectric constant of water at optical frequencies.

6. Calcite has $n_{\text{slow}} = 1.66$ and $n_{\text{fast}} = 1.49$ at $\lambda = 0.59$ μm. Find the difference in refraction angle for an incident 45-degree ray.

PROBLEMS

The graph shows the index of refraction of a material which fills all space from $z = 0$ to $z = L$. All light falling on this material may be described by

$$\mathbf{E} = (\hat{\mathbf{i}} + \hat{\mathbf{j}})E_0 \sin\left[2\pi\frac{z}{\lambda} - 2\pi\nu\mathbf{t} + \frac{7\pi}{6}\right].$$

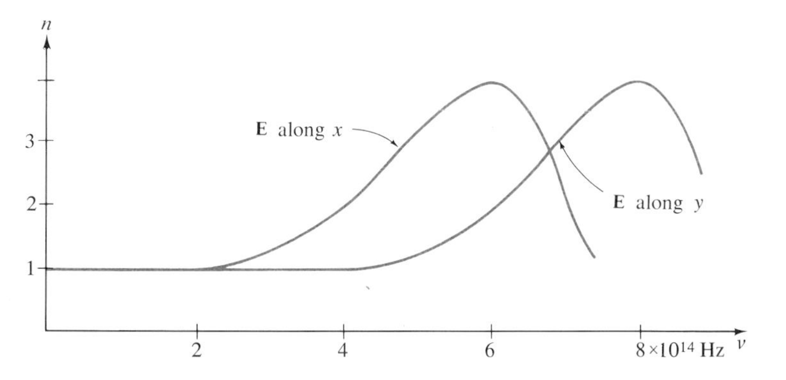

5.1 What is $\mathbf{B}(0, 0)$?

5.2 The linear momentum per unit volume at $z = 0$ is $\mathbf{p}(0, t)$. Find $\mathbf{p}(0, t)$ and its average value, $\langle\mathbf{p}\rangle_{\text{av}}$.

5.3 Find the energy density, $\mathscr{E}(z, 0)$ and its average, $\langle \mathscr{E} \rangle_{av}$.

5.4 Sketch the vector and phasor(s) for $\mathbf{E}[-(\lambda/2), 0]$ in their proper planes.

5.5 Sketch the vector and phasor(s) for $\mathbf{B}(-\lambda, \tau)$ in their proper planes.

5.6 What is the wavelength for the x component of the wave at $\nu = 6 \times 10^{14}$ Hz for $z = L/2$?

5.7 What is the wavelength for the y component of the wave at $\nu = 6 \times 10^{14}$ Hz for $z = L/2$?

5.8 Over what range of frequencies is the material fairly transparent? Where is it probably dichroic?

5.9 Find the group velocity for the y component at $\nu = 6 \times 10^{14}$ Hz.

5.10 In what range for the x component is the group velocity less than the phase velocity?

5.11 What is the energy density at $z = L/2$?

5.12 What is the linear momentum per unit volume at $z = L/2$?

5.13 What is the phase of $E_x(L, 0)$ relative to $E_y(L, 0)$ for light of frequency 4×10^{14} Hz? Take $L = \frac{5}{32} \times 10^{-6}$ m.

5.14 What is the phase of $E_x(L, 0)$ relative to $E_y(L, 0)$ for light of frequency 6×10^{14} Hz? Take $L = \frac{5}{32} \times 10^{-6}$ m.

6 Polarized light

One of the quantities needed to describe a wave is its *polarization*. That is, we must specify the direction in which the string is displaced, or in which the **E** vector points. The wave we usually describe on a string is a *transverse* wave, with the displacement at right angles to the direction in which the wave travels. The wave may travel toward $+z$, but the string may move toward $\pm x$ or $\pm y$. The same is true of an electromagnetic wave: If the wave travels toward $+z$, **E** lies in the x, y plane at right angles to the direction of propagation. If we choose a certain direction in this plane (say, $\mathbf{E} = \hat{\mathbf{j}}E$), we have chosen a polarization. Two such polarizations are possible for any transverse wave. A *longitudinal* wave, such as that of sound, can be polarized only along the propagation direction.

6.1 *Linear and circular polarization*

The transverse waves we have discussed so far have been *linearly* polarized, or *plane*-polarized. That is, the **E** vector lies along some line. Its size and sign changes, but not its direction. The same is true for the displacement vector for a particle of the string. Such a line defines, with the propagation direction, the plane of the polarization. Of course the polarization direction need not be an axis of the coordinate system. For instance, the **E** vector may lie along the line

60 degrees clockwise from the x axis, in the x, y plane: If we want to resolve this vector along some more convenient axes, we find that its components, $E_x(z, t)$ and $E_y(z, t)$, also represent linearly polarized waves. Since E_x reaches its maximum value at the same time as E_y, the components are *in phase*. Thus, two linearly polarized waves, in phase, add up to a single linearly polarized wave.

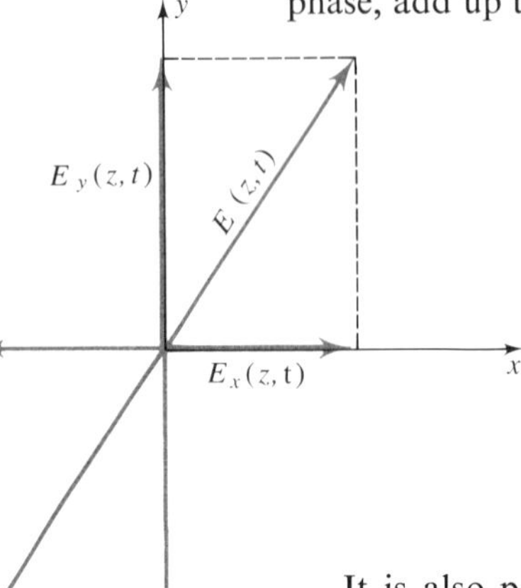

Figure 6.1: Linearly polarized light; components are also linearly polarized.

It is also possible to add two linearly polarized waves that are not in phase. In general, the resultant is an elliptically polarized wave, but we will consider a special case, that of circular polarization.

If $E_x(z, t)$ and $E_y(z, t)$ have the same amplitude but differ in phase by $\pi/2$, we get the situation shown in Figure 6.2. It is seen that

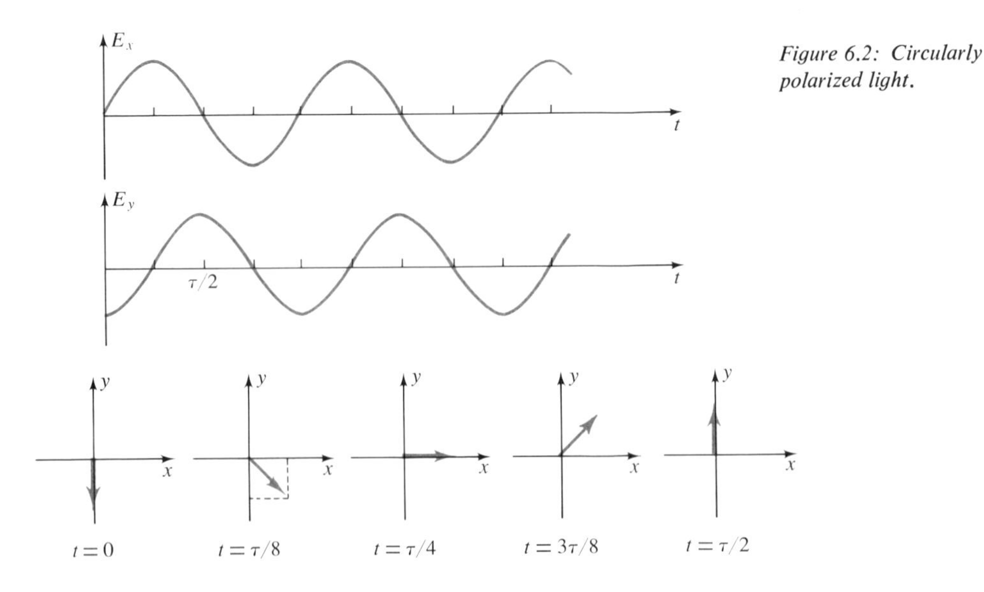

Figure 6.2: Circularly polarized light.

the end of $\mathbf{E}(z_0, t) = \mathbf{\hat{i}}E_x(z_0, t) + \mathbf{\hat{j}}E_y(z_0, t)$ traces out a circle of radius $|E| = |E_x| = |E_y|$. If this vector describes the position of the string at $z = z_0$, then that particle of string is actually going around in a circle. In fact we can generate such a *circularly polarized* wave on a string by swinging one end of it in a circle. A jump rope is a circularly polarized standing wave on a rope. (It may be operated with one or more nodes if the frequency is raised.)

Figure 6.3: Circularly polarized standing wave on a rope.

A circularly polarized electromagnetic wave is constructed in the same way, with $\mathbf{E}(z, t)$ meaning the electric-field vector. In the snapshot view of the circularly polarized wave, we see $\mathbf{E}(z, t_0)$: The string is in a helical configuration, or the electric-field vectors trace out a helix as shown in Figure 6.4.

The more complicated wave, elliptically polarized, occurs when either $|E_x| \neq |E_y|$ or the components differ in phase by other than multiples of $\pi/2$. In *all* cases, we can resolve the wave into its linearly polarized components. (6.1–6.4)

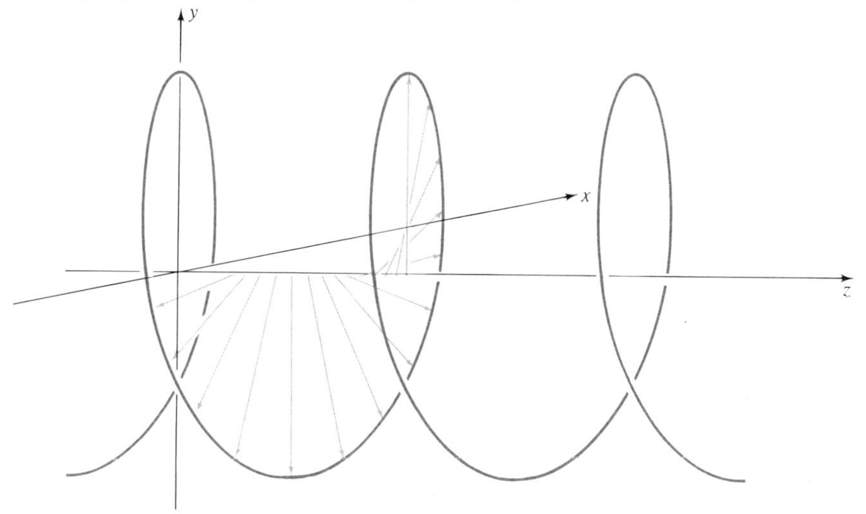

Figure 6.4: Circularly polarized electromagnetic wave vector, or string displacement.

6.2 *Production and analysis of linearly polarized light*

Let us consider the most familiar way of producing polarized light. We start with light from a common source such as an incandescent bulb or a gaseous discharge tube. This light is *randomly* polarized; that is, it includes waves that have any of the infinitely many possible transverse orientations for their **E** vectors. No matter what the orientation of a given **E** vector, we can decompose it into components along two rectangular axes (say, x and y). Now let it pass through a sheet of dichroic material, oriented in such a way that light with **E** along the x axis is absorbed, while that with **E** along the y axis is transmitted. What emerges is that component of the original wave which had the y polarization.

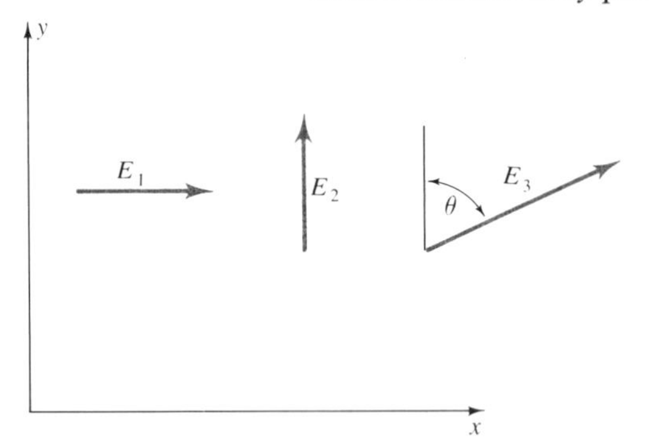

Figure 6.5: E_1 and one component of E_3 are absorbed by the analyzer.

In Figure 6.5 the wave with $\mathbf{E}_1$ will be totally absorbed. Wave 2 is transmitted without modification, but only part of wave 3 gets through. If we decompose wave 3 into two linearly polarized waves, in phase with each other, one of these components gets through, but not the other. Thus the emerging component is a wave linearly polarized in the y direction, with $E = (E_3)_y = E_3 \cos\theta$. Since the energy of the wave is proportional to E^2, the energy of the transmitted wave

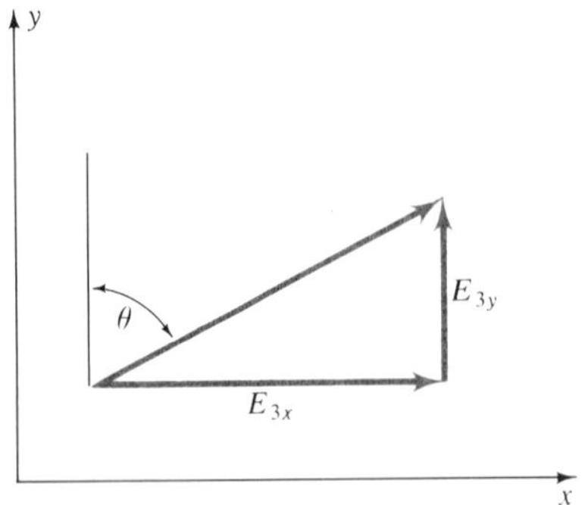

Figure 6.6: Components of E_3.

is reduced by the factor $\cos^2 \theta$. That this general result includes waves 1 and 2 is seen by inserting the values of 0 and $\pi/2$ for θ, yielding the special results of these waves as shown above. What we have done is to turn a randomly polarized (commonly called "unpolarized") beam of light into one that is polarized along the y axis. It has lost just half its original energy, as we can see from the fact that the factor of $\cos^2 \theta$, when averaged over all possible values of θ, becomes $\frac{1}{2}$. There are other ways of polarizing a beam of light, and we will look at them shortly, but first let us see how the polarized beam behaves in various situations.

The device we used to convert a randomly polarized beam of light into a polarized one is called the *polarizer*. An identical device, called the *analyzer*, may be used to test the polarization state of the beam, by seeing whether the beam can pass when the analyzer is in a given orientation. If the transmission axes of polarizer and analyzer are at right angles, they are said to be crossed, and no light is transmitted. If the two axes are at some general angle, θ, the intensity of the emergent light is given by $I = I_0 \cos^2 \theta$. This follows from our discussion above, as does the fact that I_0 is half the intensity in the original unpolarized beam. Now let us insert various substances to be investigated between the polarizer and analyzer. Surely the simplest material should be another sheet of polarizer. But look what happens! We start with the polarizer and analyzer crossed so that no light comes through. Now we insert an intermediate polarizer at an angle θ to the first polarizer. Surprisingly, now light does come through! The explanation is actually quite simple. After the first polarizer, all light waves have **E** lying along one axis, say, x. From each of these waves, the second polarizer selects the component which lies along its transmission axis and which has magnitude $E_0 \cos \theta$. Now the analyzer makes no distinction between this component and any other E along the same axis, so in turn it selects a component $(E_0 \cos \theta) \sin \theta$, which need not be zero. The corresponding intensity is proportional to the square of this: $I_0 \cos^2\theta$

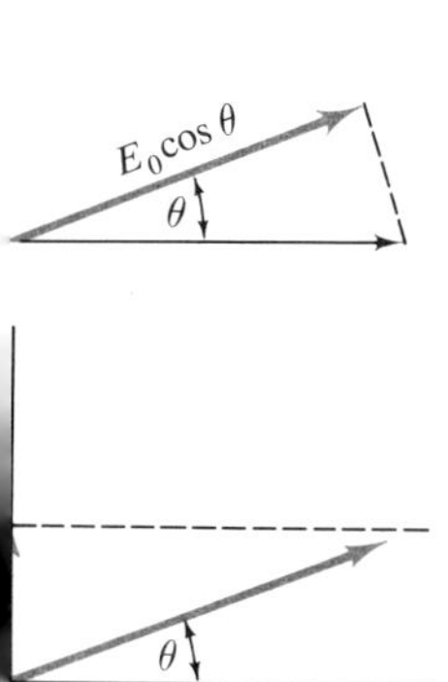

Figure 6.7: Crossed polarizer and analyzer with intervening polarizing sheet.

$\sin^2 \theta$, and is zero only if $\theta = 0$ or $\pi/2$ (that is, if the inserted piece lines up with the axis of either the first polarizer or the analyzer). If there is a lesson to be learned from this example, it is that we must keep sight of the component E vectors, since they are the ones that are involved in polarization. The advice also turns out to be valid for most of our subsequent investigations.

Now let us put a different device between polarizer and analyzer. This time we use a birefringent material. Consider a thin slab cut so that both a slow and a fast axis lie in its plane. If the light has **E** along one of these axes, it is unaffected by the other, and the slab may as well have no birefringence. But if **E** is at an angle to the axes, we must again consider its components along each axis. Since one component wave travels faster than the other, the two waves will get out of phase. In general, they emerge with different amplitudes and some unspecified phase relationship so that the total wave is elliptically polarized. This is hard to analyze and not very useful, so we will consider a few special cases, which end up with linear or circular polarizations. (6.5a)

6.3 *Wave plates*

First suppose that the phase difference is $2m\pi$, where m is an integer. In this case we get back our original polarization. Such a slab is called a *full-wave plate*. How could we design one? A phase difference of 2π means that one wave component has traveled a whole wavelength farther than the other. How can this be? Remember that the wave with its E vector along the fast axis, f, travels with speed $v_w = c/n_{\text{fast}}$, which is greater than the speed of the other component

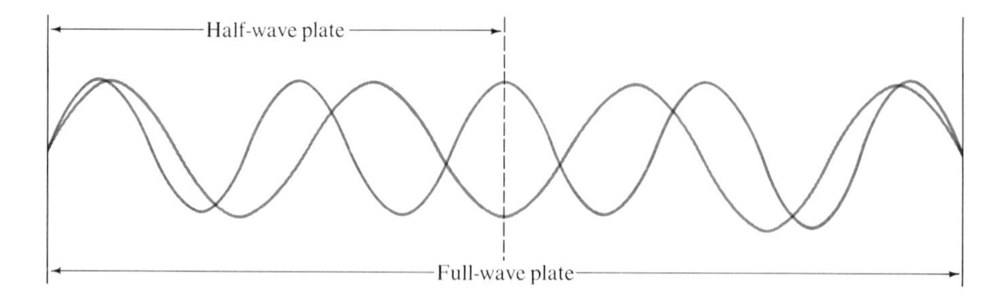

Figure 6.8: Half- and full-wave plates.

wave. Since both waves have the same frequency, the faster one has a longer wavelength. If the slab has a thickness L, such that $L = 91\lambda_{\text{fast}} = 93\lambda_{\text{slow}}$, then the slow wave has gone two wavelengths farther than the fast wave (equivalent to a phase difference of 4π), but they emerge in phase and so appear unchanged. A bit of arithmetic shows that the proper thickness for a full-wave plate, then, is $(n_{\text{slow}} - n_{\text{fast}})L = m\lambda$, where m is an integer and λ means the wavelength in vacuum.

The discussion above leads to an obvious definition of a *half-wave plate*: $(n_{\text{slow}} - n_{\text{fast}})L = (2m + 1)\lambda/2$. This means that if the two components go into the slab in phase, they emerge from the slab 180 degrees out of phase. What effect will this have? As before, if **E** lies along an axis, there is no effect; so let us consider a wave with **E** at the general angle θ to, say, the fast axis. We take components of the vector **E** along these axes. The emergent wave has components of the same size as the incident one, but they are now out of phase. A *relative* phase change of π is equivalent to reversing one

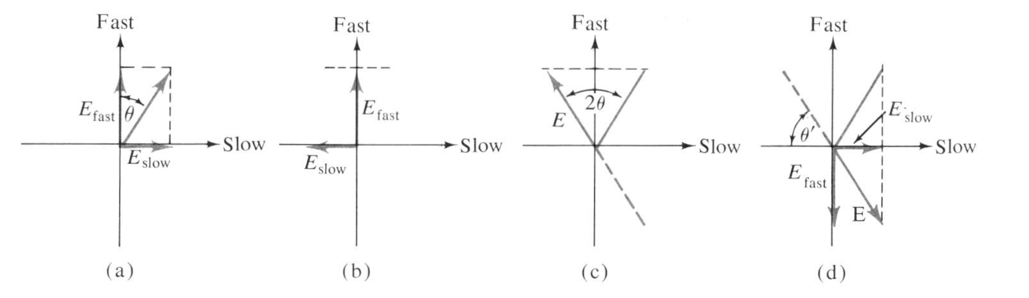

Figure 6.9: Rotation of polarization by a half-wave plate. (a) Vector before passing through half-wave plate. (b) After half-wave plate, E_{slow} sin α has become $E_{\text{slow}} \sin(\alpha + \pi) = -E_{\text{slow}} \sin \alpha$. (c) The resultant vector has rotated through 2θ. (d) If E_{fast} had been reversed, the result would have been the same.

of the vectors. By recombining, we find that the polarization direction has been rotated through twice the angle θ. Note that this is independent of which component we chose to reverse, since the polarization direction only specifies the line along which **E** oscillates. Thus, the half-wave plate rotates the polarization direction, but otherwise leaves the wave unchanged. It does this by changing by 180 degrees the *relative* phases of the fast and slow components of **E**.

An alternative way of looking at this example is this: We think of the vector **E** as the resultant of two components E_{slow} and E_{fast}, and of these, in turn, as the resultants of components along axes parallel and perpendicular to **E**, as in Figure 6.10. Of the four ultimate

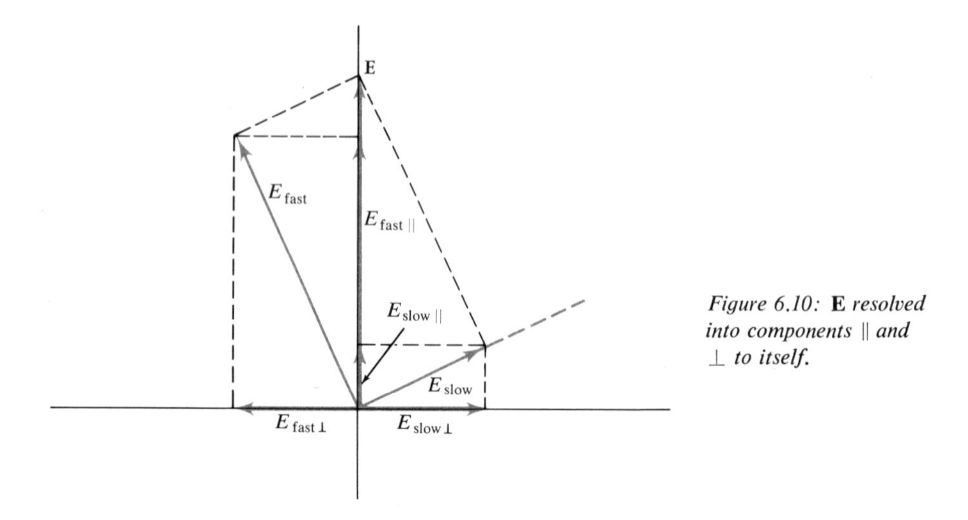

Figure 6.10: **E** *resolved into components* || *and* ⊥ *to itself.*

components, $E_{fast\|}$ and $E_{slow\|}$ are collinear and in phase, and add up to **E**. We can also add $E_{fast\perp}$ and $E_{slow\perp}$, since they are collinear, and we see that they cancel out. Now the half-way plate changes by 180 degrees the *relative* phase of the slow components so that $E_{slow\perp}$ and $E_{slow\|}$ are reversed in direction. Recombination then gives the same result as that found above.

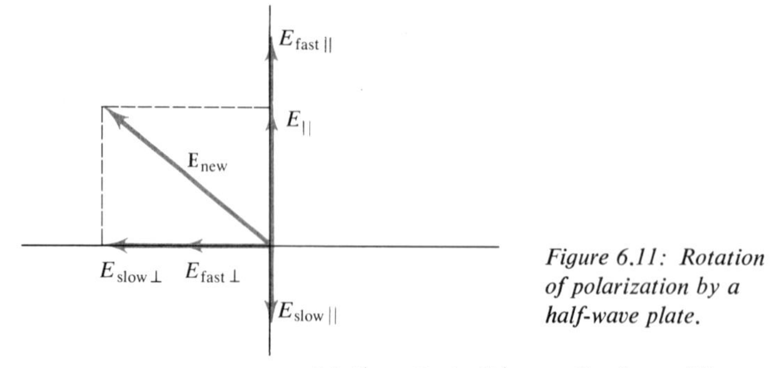

Figure 6.11: Rotation of polarization by a half-wave plate.

Notice that this method readily explains the three-polarizer result: The first polarizer selects **E**, the second eliminates *both* $E_{slow\|}$ and $E_{slow\perp}$, and the third (the analyzer) eliminates $E_{fast\|}$, leaving $E_{fast\perp}$. (6.5b; 6.6–6.8)

6.4 *Colors* We often see colors appearing when a birefringent material is placed between polarizer and analyzer, even though each piece alone is color-neutral. The reason for this is that a birefringent slab will be an nth wave plate at one frequency only. For instance, a half-wave plate for red light $\{L(n_{\text{slow}} - n_{\text{fast}}) = \frac{1}{2}(760\text{ nm})\}$ will be a full-wave plate for deep-violet light $\{L(n_{\text{slow}} - n_{\text{fast}}) = 1 \cdot (380\text{ nm})\}$. Thus, if we put the slab between crossed polarizer and analyzer, with axes at 45 degrees, no violet light comes through the system. But the **E** vector for red light is rotated 90 degrees so that it *does* pass the analyzer. For frequencies in between, the light is elliptically polarized and must be treated by components, but we can see that less and less light comes through as we go from red to violet. Thus, looking through the system at white light, we see light of a red-orange color. If we turn the analyzer 90 degrees, the violet comes through, but not the red, so that we get the complementary colors.

Good commercial wave plates are made for the center of the spectrum and may have some compensation, due to n_{slow} being a slightly different function of λ than is n_{fast}. But various sheet plastics such as those used in packaging and in mending tapes show strong color effects. Repeated layers of cellophane, for example, make the colors deeper (more saturated) as the departure from the nth wave plate is built up. For instance, if $L(n_{\text{slow}} - n_{\text{fast}}) = 500$ nm (nanometers), we have a full-wave plate at 500 nm and nearly a full-wave plate at 456 nm. But five layers of this leave us with a full-wave plate at 500 nm and a half-wave plate at 456 nm $(5L(n_{\text{slow}} - n_{\text{fast}}) = 5.5 \times 456\text{ nm})$.

The reason many plastics are birefringent is that they have been stretched in manufacture so that their long molecules line up somewhat. This also happens when transparent materials are strained; the resultant colors are useful in locating and analyzing such strains. Typically, rotation of any part of the system results in the complementary colors.

6.5 *Circularly polarized light* It should be clear that a *quarter-wave plate* will be one that has a thickness $(n_{\text{slow}} - n_{\text{fast}})L = (2m + 1)\lambda/4$, and which changes the relative phases of the fast and slow components of **E** by 90 degrees. In general, the result of such a phase change will be elliptical polarization. Linear polarization is a special case of elliptical polarization and occurs when **E** lies along an axis, as we already know. The other special case is circular polarization. This is obtained from linearly polarized light by passing it through a quarter-wave plate whose axes are at 45 degrees to the **E** vector of the wave.

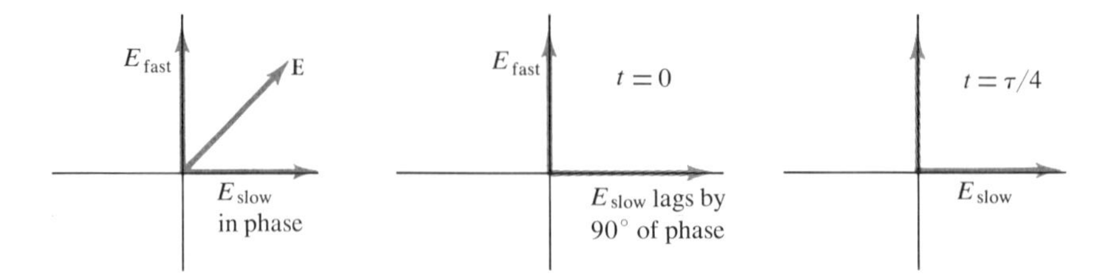

Figure 6.12: Circular polarization.

There are two possible circular polarizations, of course, since the **E** vector can rotate to the right or left, just as the jump rope could. We get one or the other according to whether **E** is 45 degrees clockwise or counterclockwise from the fast axis. Of course, although it is easy to identify the axes, it is not easy to find which is fast and which is slow. Thus, while we can easily change from right to left circular polarization, or vice versa, we may not know which is which. In fact, it seldom matters, since we are more concerned with differences in polarization.

We have said that the randomly polarized light from most sources is made up of light linearly polarized along two perpendicular axes (which we select quite arbitrarily). We might equally well have said that such unpolarized light consists of random combinations of right and left circularly polarized waves. Just as two linearly polarized waves may be added to form a circularly polarized one, we can add two circularly polarized waves to form a linearly polarized one. (6.5c; 6.9–6.12)

6.6 *Angular momentum of light*

Circularly polarized light waves carry angular momentum, a fact that is significant in many applications of quantum mechanics. For instance, when an atom emits or absorbs a photon, it loses or gains angular momentum. This involves the angular momentum of light in the "selection rules" which play such an important role in modern physics. We can calculate the magnitude of the angular momentum by a procedure that is much the same as the one used to calculate the linear momentum: We arrange for some charges to absorb the light and calculate how much angular momentum they have gained. For simplicity of visualization, we will think of "free" charges in a metal.

The electric field of the wave, **E**, describes a circle during each cycle; so the charges will be accelerated in the same way as a satellite accelerates in the earth's gravitational field. The result is that the charge follows a circular path, as shown in Figure 6.13.

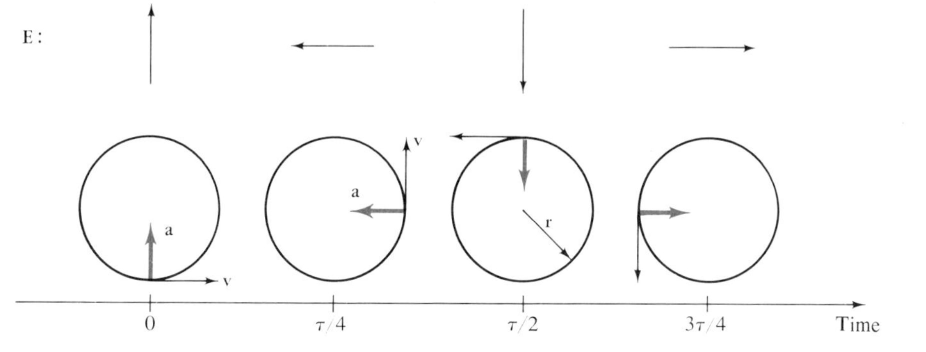

Figure 6.13: A charge q under the action of a circularly polarized electromagnetic wave.

If the radius of this path is r, we can calculate the torque on the charge, $\mathbf{\Gamma} = \mathbf{F} \times \mathbf{r} = \hat{\mathbf{k}} qEr \sin\theta$, where θ is the angle between **F** and **r**. Similarly, we can measure the power absorbed (the work done on the charge) from the wave: $P_{\text{abs}} = \mathbf{F} \cdot \mathbf{v} = qEr\omega \sin\theta$, where $\omega = 2\pi\nu$ is the angular velocity.

Here, θ is not equal to zero because some friction is present. This means that **F** must have a component *along* the direction of **v**. This is shown in Figure 6.14, with the angle exaggerated. Notice that we

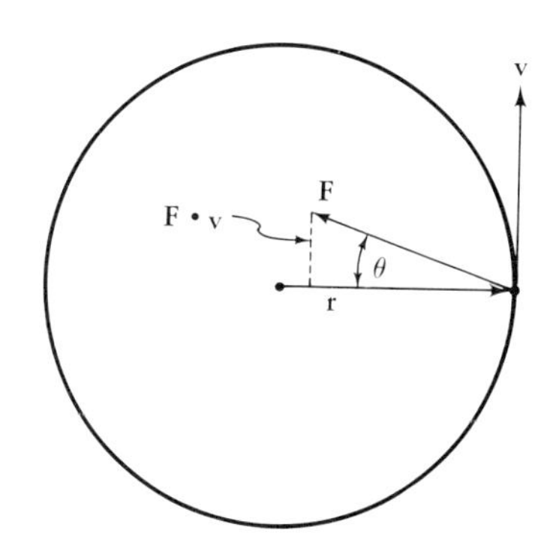

Figure 6.14: Forces on a charge due to a circularly polarized wave.

have here the same situation that was encountered in the calculation of the linear momentum: If there are no dissipative forces, no power will be absorbed.

Again, we could relate Γ and P_{abs}, but it is more useful to use the relations $\Gamma = d\mathbf{l}/dt$, where $\mathbf{l}$ is the angular momentum, and $P_{abs} = dU/dt$. Then $l = U/2\pi\nu$ when all the energy in the wave has been absorbed. (6.14–6.16)

Angular momentum is a vector quantity, so we must consider its direction as well. The charge moves in the plane containing **E**, and therefore moves in the plane perpendicular to the propagation direction. So the angular momentum will be either along the propagation direction or in the opposite direction, depending on which way **E** was rotating. Since the linear momentum is *along* the propagation direction, we can use it to *define* right circular polarization: The wave is *right* circularly polarized when $\mathbf{l}$ and **p** point the same way; *left*, when they are opposed.

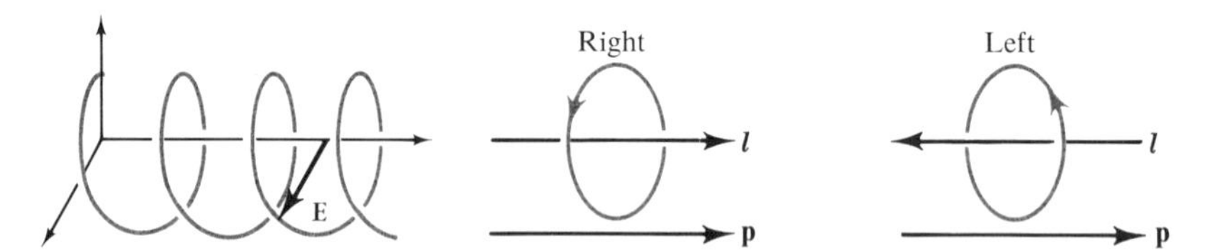

Figure 6.15: Definitions of right- and left-circular polarization for light.

When we discussed linear momentum, we found it useful to "run time backward" in order to study the reaction of an emitter as well as an absorber. Let us do the same in this case: Since **v** changes sign, but $\mathbf{F}(=q\mathbf{E})$ does not, the sign of dU/dt changes (the power is emitted). The torque is unchanged, so the electron is now slowed from $-\mathbf{v}$ to 0, leaving the mirror with no net angular momentum after the re-emission. Since the mirror has not changed its angular momentum, neither can the wave. Thus the re-emitted wave has the same angular momentum as the original one had. According to our definition, then, the polarization has been reversed because the linear momentum *has* changed sign. This is the basis for constructing a "light trap", as shown in Figure 6.16.

We have seen, then, that electromagnetic waves can carry angular momentum as well as linear momentum. Upon reflection, the linear momentum is reversed, but the angular momentum is not. This means that the (circular) polarization of the light is reversed on reflection. Linearly polarized light, which can be thought of as a superposition of two circularly polarized waves, carries zero net angular momentum. The concept of a photon needs to be considered

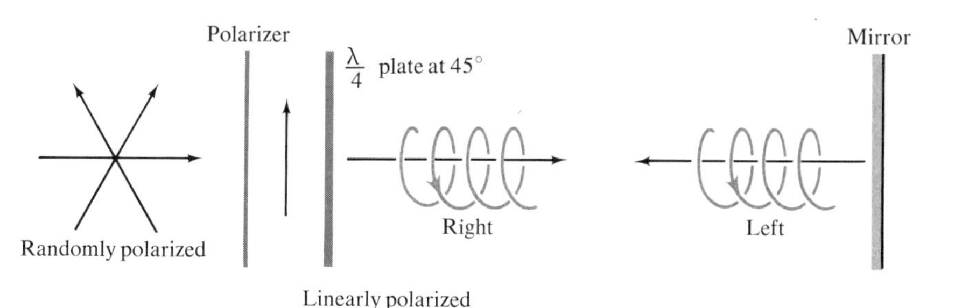

Figure 6.16: Light trap: The reflected light is left-circular polarized and cannot pass the right-circular polarizer to re-emerge.

in these same terms. We might think of the photon as linearly polarized (in one of two perpendicular directions) or as circularly polarized (in one of two rotation senses). In the linear "basis," it has no angular momentum, while in the circular one, it has an intrinsic angular momentum of $l = U/2\pi\nu = h/2\pi$. All circularly polarized photons have the same amount of angular momentum, independently of frequency.

Linearly polarized photons are thought of as a superposition of circular ones, and the converse is also true. Transverse waves generally have this variability of basis. The same apparent ambiguity arises in many situations of quantum mechanics. The most dramatic is the wave-particle dualism itself. In practice, one must choose a convenient basis in a given problem and then work within it. In changing, care must be taken not to violate some restriction imposed by the earlier basis.

6.7 *Other polarizing interactions*

We have been discussing polarized light as if only birefringent and dichroic materials affected polarization states. This is not true, although such materials are indeed the most convenient. These instances of light's interaction with matter involve anisotropically bound electric charges which scatter the wave; that is, they absorb and re-emit the wave. Even when the scatterers are not bound anisotropically, the polarization may be involved. We will look at some examples of ordinary scattering processes to see this. We will also consider briefly a few of the processes whereby light is produced with polarization already specified. All these processes depend on the following fact: When an oscillating charge gives rise to an electromagnetic wave, the **E** vector of the wave lies along the direction of oscillation. Since light waves are transverse, they may therefore travel in any direction perpendicular to the oscillation direction, but they may *not* be emitted *along* the oscillation direction. A convenient way to remember this is to note that an observer looking along the motion of the charge does not see it accelerated and detects no radiation.

Our first application of this information will be to the scattering of light by small, isolated particles. Such particles may be the molecules of a gas or dust particles in the air. Let light polarized linearly in the x direction fall on such a system. The charges are accelerated

Figure 6.17: Polarized light is not scattered along the polarization (**E**) *direction.*

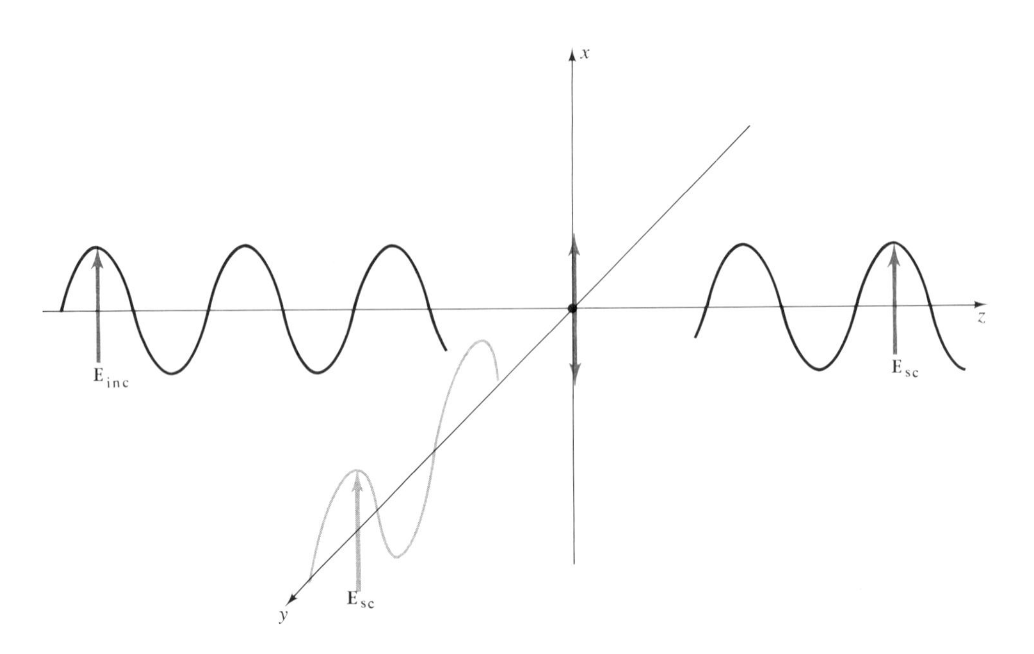

in the x direction, so the scattered waves may go only in the y and z directions. Thus, if unpolarized light falls on the gas, that scattered forward is still unpolarized, but the light scattered sideways will be polarized vertically. It is easily verified that this is true for sunlight scattered from the atmosphere. The effect is complete for light scattered at right angles and is partial at an arbitrary angle. Only by

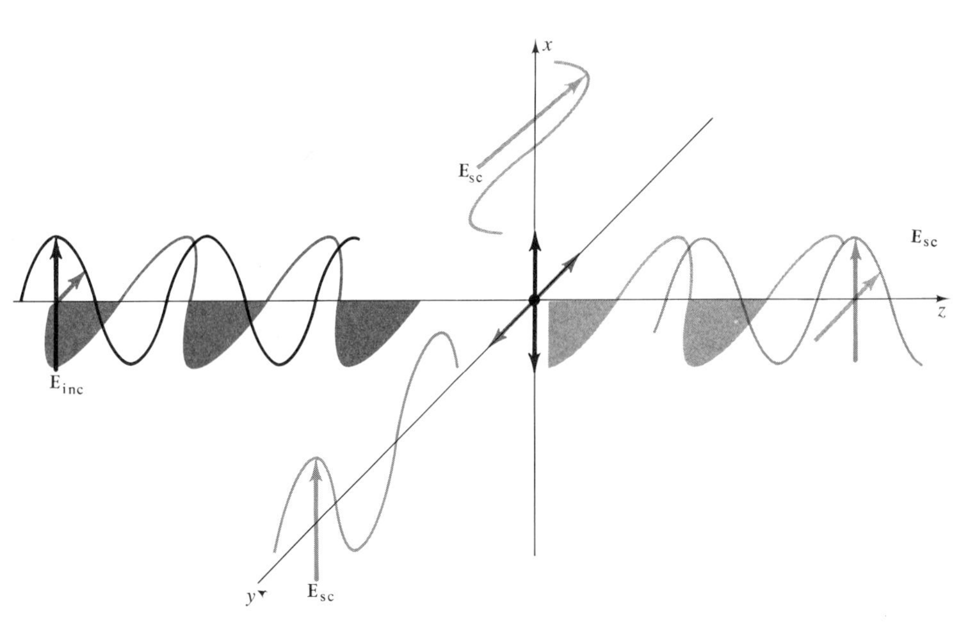

Figure 6.18: Polarization by scattering.

looking directly at the sun does one see light that is entirely unpolarized. (6.17–6.18)

Perhaps the commonest polarization effect is that of reflection from nonmetallic surfaces. To be reflected from the surface of glass, light must be scattered by the charges bound in the region near the surface. In the bulk material, such side scattering is imperceptible, due to the randomizing of the phases of the scattered wavelets, but near the surface it is significant.

We have two restraints on the reflection: The angles of incidence and reflection must be equal, and no light may be scattered along the

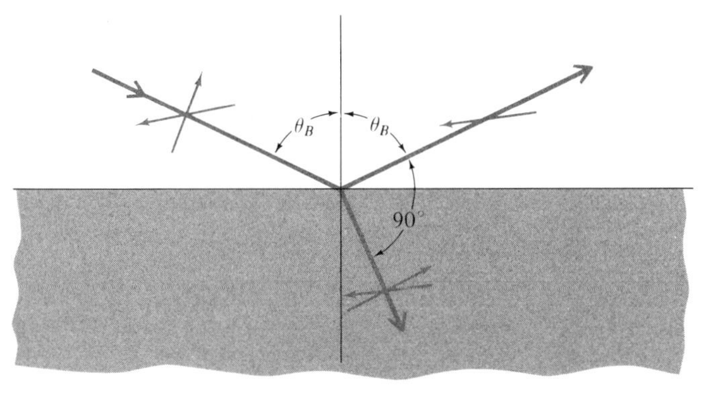

Figure 6.19: Polarization by reflection.

motion of the charge. Thus at "Brewster's angle" θ_B, only light polarized with **E** in the plane of the surface will be reflected. This angle is the one for which the reflected ray makes a 90-degree angle with the *refracted* one, since it is inside the glass that the scattering occurs. Using the laws of reflection and refraction, we find that tan $\theta_B = n$, where n is the index of refraction of the material. This provides a way of measuring the indices of opaque materials. We can also use this mechanism to convert unpolarized light into linearly polarized light. Near Brewster's angle, the effect is noticeable but incomplete.

A result similar to dichroism may be achieved for microwaves by letting them fall on a set of parallel wires. If **E** is along the wires, currents are induced, ohmic heating takes place, and the wave is absorbed. If, however, **E** is perpendicular to the wires, the charges are "bound" to the width of the wire, and so reradiate the wave like the charges in an insulator. Such polarizers behave in all ways like optical sheet polarizers.

Radio waves (including microwaves) come ready-polarized because they are generated by charges constrained to oscillate along the line of an antenna. Reciprocally, the antenna must be along **E** in a receiver. (Sometimes a "loop" antenna is used, in which case its plane must be perpendicular to **B**.) A portable radio demonstrates this easily: turn it 90 degrees and blessed silence ensues. A few of the sources of electromagnetic waves in nature give rise to polarized emission. Such a situation requires that something select a direction, such as the magnetic fields associated with the polarized radiation from sunspots and the Crab nebula. A recent source of polarized light, the laser, uses the same scattering that we have seen before, with the addition of energy at each re-emission. Since scattering does not change the polarization, the energetic emergent light has the same polarization as the first weak wave to be scattered and amplified. This oversimplifies a complicated process, but the result (a polarized beam) is legitimately explained in this way.

In all our considerations of polarization effects, we have found it useful to regard the interaction of electromagnetic waves with matter as a scattering process. This is a generally useful approach. Light is most interesting when it is interacting with matter, and this must, of necessity, mean scattering from charges. Extension to quantum mechanical effects does not lessen this observation, and many sophisticated interactions are best considered first in this simple way.

EXERCISES

1. What is the polarization of sea waves? Of broadcast radio waves?

2. Unpolarized light of intensity I_0 falls on three polarizing sheets with axes at 0, 30, and 60 degrees. What is the final light intensity?

3. Use data from a handbook to find how thick a quartz quarter-wave plate should be.

4. Find the thickness of a half-wave plate made of calcspar, using handbook data.

5. A 50-kW radio station has a frequency of 10^6 Hz. If its output is circularly polarized, what is the torque on its antenna?

6. Steady current in a solenoid is interrupted, and a concentric solenoid with ends shorted starts to turn. Explain.

PROBLEMS

6.1 Describe the polarization state of the following waves:

(a) $\mathbf{E}_T = \hat{\imath}E_0 \sin[2\pi(z/\lambda - \nu t)] + \hat{\jmath}E_0[\cos 2\pi(z/\lambda - \nu t)]$.

(b) $\mathbf{E}_T = \hat{\imath}E_0 \sin[2\pi(z/\lambda + \nu t)] + \hat{\jmath}E_0 \sin[2\pi(z/\lambda + \nu t - \frac{1}{8})]$.

(c) $\mathbf{E}_T = \hat{\imath}E_0 \sin[2\pi(z/\lambda - \nu t)] - \hat{\jmath}E_0 \sin[2\pi(z/\lambda - \nu t)]$.

6.2 Describe the polarization state of the following waves:

(a) $\mathbf{E}_T = \hat{\imath}E_0 \cos[2\pi(z/\lambda - \nu t + \frac{1}{6})] + \hat{\jmath}E_0 \sin[2\pi(z/\lambda - \nu t + \frac{1}{6})]$.

(b) $\mathbf{E}_T = \hat{\imath}E_0 \cos[2\pi(z/\lambda + \nu t)] + \hat{\jmath}E_0 \cos[2\pi(z/\lambda + \nu t - \frac{1}{8})]$.

(c) $\mathbf{E}_T = \hat{\imath} \exp[-(z/\lambda - \nu t)^2] + \hat{\jmath} \exp[-(z/\lambda - \nu t)^2]$.

6.3 Even elliptically polarized waves can be made up of linearly polarized components. This allows us to work with much more complicated waves without acknowledging their complexity until the final result. For instance:

(a) Two linearly polarized waves are in phase, but have different amplitudes. At $x = 0$,

$$\mathbf{E}_1 = \hat{\imath}A_1 \cos 2\pi\nu t + \hat{\jmath}B_1 \cos 2\pi\nu t,$$

$$\mathbf{E}_2 = \hat{\imath}A_2 \cos 2\pi\nu t + \hat{\jmath}B_2 \cos 2\pi\nu t.$$

Show that $\mathbf{E}_T$ is also linearly polarized, and find its polarization direction.

(b) Two circularly polarized waves (a right and a left) can be added to form a linearly polarized wave:

$$\mathbf{E}_1 = \hat{\imath}E_0 \cos 2\pi\nu t + \hat{\jmath}E_0 \sin 2\pi\nu t,$$

$$\mathbf{E}_2 = \hat{\imath}E_0 \cos(2\pi\nu t + \alpha) - \hat{\jmath}E_0 \sin(2\pi\nu t + \alpha),$$

$$\mathbf{E}_T = \mathbf{E}_1 + \mathbf{E}_2.$$

Show that $\mathbf{E}_T$ is linearly polarized, and find its polarization direction.

(c) An elliptically polarized wave is written (at $z = 0$) as

$$\mathbf{E}_T = \hat{\imath}A \sin 2\pi\nu t + \hat{\jmath}B \cos 2\pi\nu t.$$

Show that this can be decomposed into a linearly and a circularly polarized wave.

6.4 Show that the elliptically polarized wave

$$\mathbf{E}_T = \hat{\imath}A \sin\theta + \hat{\jmath}B \sin(\theta + \beta), \qquad \theta = 2\pi(z/\lambda - \nu t)$$

is equivalent to two circularly polarized waves, of amplitudes

$$\frac{A^2 + B^2 - 2AB\sin\beta}{4} \quad \text{and} \quad \frac{A^2 + B^2 + 2AB\sin\beta}{4}$$

6.5 Light, randomly polarized to begin with, has initial intensity I_i. It goes through three devices in sequence and emerges with final intensity I_f.

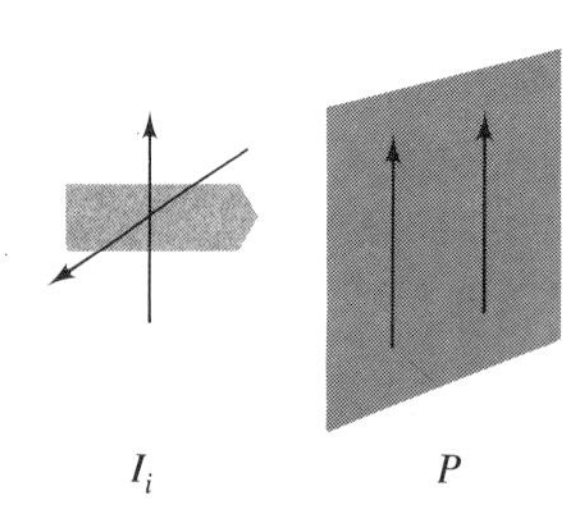

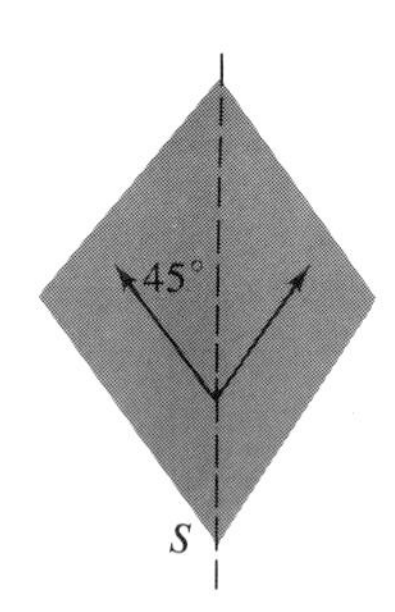

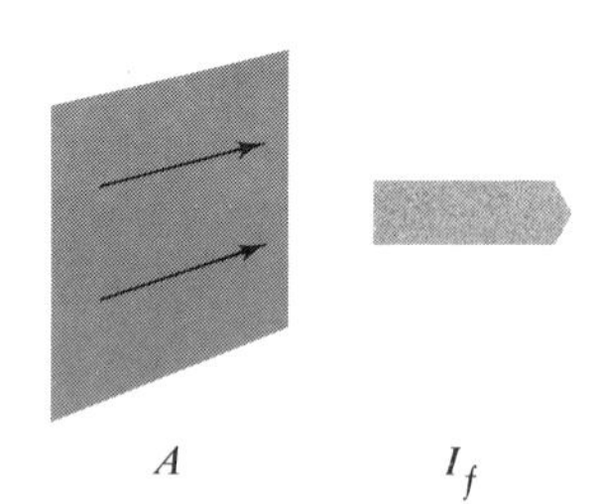

The polarizer P is a sheet of linear polarizer with polarization axis vertical. S is a sample under investigation, always with its axes, if any, at 45 degrees to the vertical. The analyzer A is a sheet of linear polarizer with axis horizontal. Find I_f/I_i when S is a

(a) Sheet of linear polarizer.

(b) Half-wave plate.

(c) Quarter-wave plate.

6.6 A number $(2N + 1)$ of sheets of linear polarizer are stacked together. Each has its axis rotated by $\pi/2N$ with respect to the previous one. What is the intensity of the emergent light, relative to the incident light?

6.7 A number $(2N + 1)$ of sheets of material are stacked together, each with axis rotated by $\pi/4N$ with respect to the previous one. The first and last are polarizers; the remainder are half-wave plates. Find the ratio of final to incident light intensities.

6.8 For a half-wave plate between crossed polarizer and analyzer, find the intensity of emergent light as a function of the angle θ between the fast axis and the polarizer axis. (Refer to Figure 6.10.)

6.9 For a quarter-wave plate between crossed polarizer and analyzer, find the intensity of emergent light as a function of the angle θ between the fast axis and the polarizer axis. (Refer to Figure 6.10.)

6.10 Does the circularly polarized wave have a node or an antinode at the reflector, for **E**? for **B**? Explain.

6.11 Many organic materials exhibit "optical activity". This means that they rotate the plane of polarized light by some angle α per unit length as it passes through them.* Find the angle at which a half-wave plate must be set to compensate for the rotation of 10 cm of a cane sugar (sucrose) dilute solution in water.

6.12 Can you identify right or left circular polarization if you know the sign of the optical activity? Explain.

6.13 Show that if **v** in the first part of Figure 6.13 is initially reversed, the charge will still end up rotating in the direction indicated. (This is *not* the same as reversing t, which reverses the rotation sense of **E** as well.)

6.14 Rewrite the expressions for P_{abs} and Γ, expressing the frictional effects as a phase lag δ rather than as a geometric angle θ.

6.15 An unpolarized beam of light of wavelength λ and power P falls for 5 sec on a linear polarizer (a), then on a quarter-wave plate (b) with axes at 45 degrees to the axis of (a), and then on an absorber (c). If all were initially at rest, find their final linear and angular momenta. Give directions where possible.

P

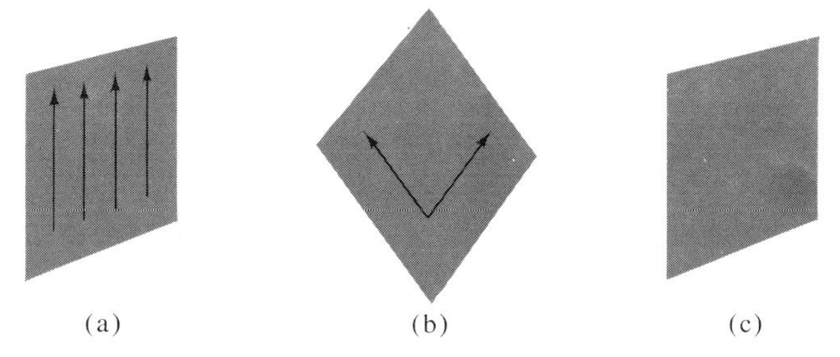

* See D. Sayers and R. Eustace, "The Documents in the Case." Avon New York, 1968.

6.16 The absorber (c) in Problem 6.14 is replaced by a reflector. Find the final linear and angular momenta.

6.17 A primitive spider web is shown in the figure. A fly twitches string R. Find the amplitude of the wave the spider detects on strings S, T, and U if the twitch is vertical; if horizontal.

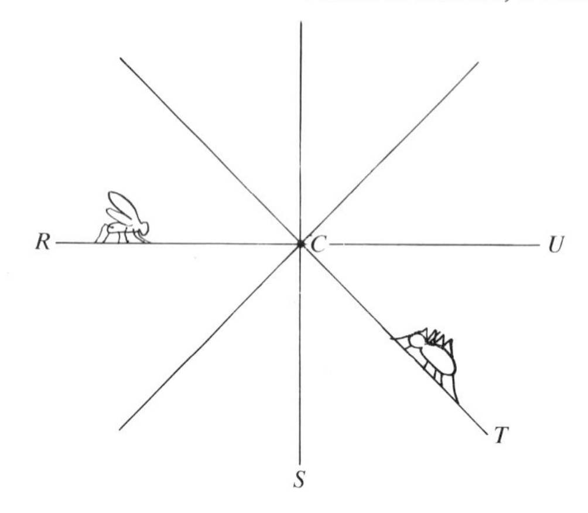

6.18 Acoustic waves are polarized longitudinally. If the atoms in a wall act like point-scatterers connected by springs to fixed centers, find the variation of intensity with angle for the reflected wave.

6.19 Prove that a linear polarizer transmits half the energy of an unpolarized light beam.

7
Interference

In this and the next few chapters we will be studying the phenomenon of *interference*. Two waves are said to interfere when we observe regions in which the total energy is not the sum of the energies of the two waves. The causes of interference involve concepts that we have already studied, such as superposition. In a sense, then, interference phenomena constitute the applications for which we developed wave optics. It is true, however, that the interference effects of optics correspond closely to those encountered in quantum mechanics, where the waves involved are the more complicated waves of the material particles. The profound importance of these interference effects makes their study in optics all the more pertinent.

7.1 *Two identical sources—in line*

Consider a situation in which we have two sources of waves, which are identical in every detail and which lie on the z axis along which the waves move. We know that we can describe the total wave disturbance as

$$E_T(z,t) = E_0 \sin\left\{\frac{2\pi}{\lambda}[(z - z_1) - ct] + \phi_1\right\} + E_0 \sin\left\{\frac{2\pi}{\lambda}[(z - z_2) - ct] + \phi_2\right\}$$

where we have written E, although the discussion will also apply equally to any kind of wave. The points z_1 and z_2 are the source points from which the waves start, while the general point z is the observation point where E_T is measured. For simplicity we have assumed that both waves started at time $t = 0$ (and are observed at time t later).

If the sources are identical, they have identical phase. Think of the waves as existing on parallel ropes. The sources are two hands, moving up and down, and holding the ropes at positions separated by

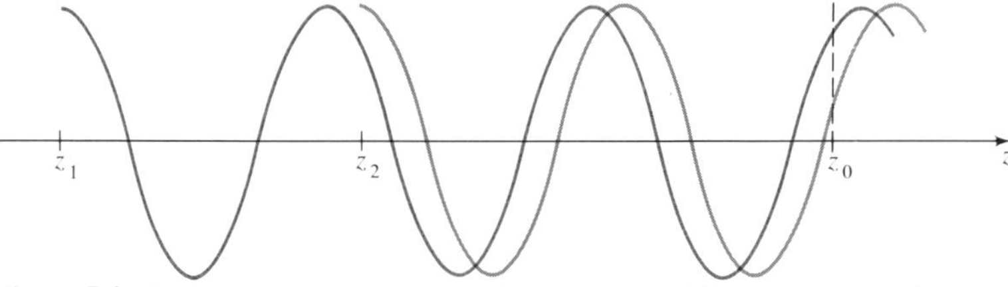

Figure 7.1: Waves from 2 sources on the z axis.

the distance $L = z_1 - z_2$. If the two hands go up and down in synchronism, the *sources* are in phase. If one hand reaches its peak position Δt seconds after the other, the sources are out of phase by an amount $\Delta\phi_{sc} = 2\pi\,\Delta t/\tau = \phi_1 - \phi_2$.

Now mark the ropes at some position z_0. If the two marks rise and fall together, we say that the *waves* are in phase at the point z_0. But this will not be the case, in general, even when $\Delta\phi_{sc} = 0$, because the sources are separated in space. This is illustrated in Figure 7.1. The waves will be in phase for certain values of the separation, L. If L is any integral number of wavelengths, crests coincide and the waves are in phase. Similarly, if the separation is half a wavelength, the two waves are said to be out of phase, and their superposition is equal to zero

A general expression for the phase difference introduced by the separation is

$$\Delta\phi = \frac{2\pi L}{\lambda} = \frac{2\pi(z_1 - z_2)}{\lambda}.$$

As usual, this is most easily checked against the known values. When

$L = 0$ the phase difference is zero, and when $L = m\lambda$, the phase difference is $m(2\pi)$, which is the same as zero. (As usual, $m = 0, \pm 1, \pm 2, \ldots$.) When $L = \lambda/2$, the phase difference is π, which means that one wave is the negative of the other, as predicted.

If a phase shift were present between the sources, it would simply add to that due to the separation:

$$\Delta\phi = \frac{2\pi L}{\lambda} + \Delta\phi_{sc}.$$

We can now write our superposed waves in a simpler form:

$$E_T = E_0 \sin\left[\frac{2\pi}{\lambda}(z - ct) + \phi\right] + E_0 \sin\left[\frac{2\pi}{\lambda}(z - ct) + \phi + \Delta\phi\right].$$

Here we have written ϕ for $\phi_1 - (2\pi/\lambda)z_1$. Since we are not interested in the actual values of these quantities, we might as well set $\phi = 0$.

Our task will be to evaluate E_T and the energy (or, equivalently, the intensity) at some observation point. This will be a function of the phase difference, or equally of the geometry leading to the phase difference. It will usually be easier to understand the general result if we look first at some special cases such as $\Delta\phi = 0$ and $\Delta\phi = \pi$. In the example above these are quite trivial: if $\Delta\phi = 0$ ($L = 0$), then the waves are in phase and add everywhere to give $E_T = 2E_0$, and therefore $\mathscr{E}_T = 4\mathscr{E}_0$. The same result, of course, pertains if $\Delta\phi = m(2\pi)$ ($L = m\lambda$), where m is any integer. If on the other hand, $\Delta\phi = \pi$ or $(m + \frac{1}{2})2\pi$ ($L = \lambda(m + \frac{1}{2})$), then the total disturbance is everywhere zero, $E_T = 0$ and $\mathscr{E}_T = 0$.

The student should verify that the general expression is

$$E_T = 2E_0 \sin\left[\frac{2\pi}{\lambda}(z - ct)\right] \cos\frac{\Delta\phi}{2}.$$

Upon squaring and taking the average over many cycles, we obtain $\mathscr{E}_T = 4\mathscr{E}_0 \cos^2(\Delta\phi/2)$. Of course, for electromagnetic waves, we will want to note that the sum is a vector one, but the sources are identical so that the two waves have the same polarizations, and we need not worry about that for the moment. Although we must keep in mind that this discussion is equally true for other waves, such as sound and water waves, we will acknowledge that our major interest is light, and will say that regions of increased intensity are *bright* and those where the intensity is reduced are *dark*. Remember that for light, the intensity is the power per unit area, and is therefore proportional to the (time averaged) energy. (7.1)

7.2 *Two identical sources—off axis*

Suppose now that our two identical sources do not lie on the z axis. The situation is no longer one-dimensional, so we must decide what sorts of waves the sources are emitting. We think of most light sources as emitting spherical waves—waves whose *intensity* falls off as $1/r^2$. Such a function for the intensity requires wave fronts that are spherical rather than plane. Emitters need not be isotropic for the waves to be spherical, as can be seen by letting only part of a spherical wave surface emerge from an aperture (a lens opening, for instance). Many of the sources we will consider are long lines of emitters from which the intensity drops off as $1/r$, and the wave fronts are cylindrical, comprising the superposition of many spherical wavelets.

All this gets fairly complicated, so we will generally take refuge in the fact that as the distance from the source increases, all wave fronts begin to approximate planes. The conditions under which the plane-wave approximations are valid may be many and diverse. When they are fulfilled, the system is said to be in the Fraunhofer limit. Let us suppose for the time being that we are indeed far enough from our sources to regard all waves as plane, and see how we should treat the problem of the two identical sources not on the line of wave travel.

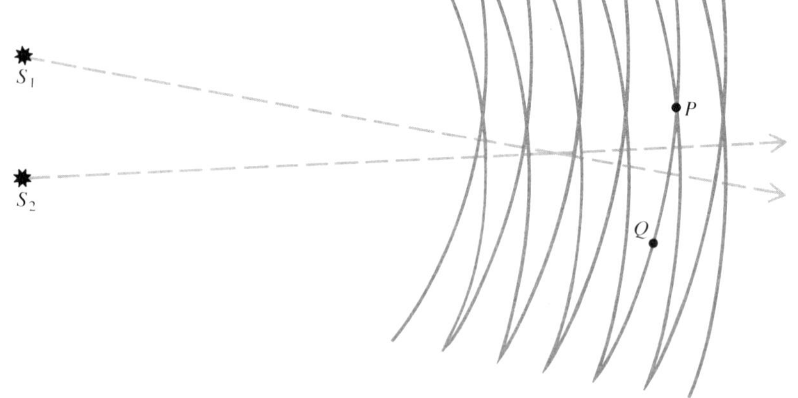

Figure 7.2: Wave fronts from 2 sources not on the propagation axis.

Figure 7.2 shows the sort of geometry with which we are concerned, where the solid lines represent wave crests. Now let us apply the principle of superposition at various points. Remember that superposition simply means that we take the sum of the two wave amplitudes to get the resultant amplitude at any place. For instance, two wave fronts coincide at point P and the resultant amplitude is $2E_0$, whereas at Q, a wave front (crest) coincides with a wave trough and the resultant is zero. The corresponding energy densities are $4\mathscr{E}_0$ at P, a bright point, and zero at Q, a dark point. The thing we will want to decide is how to locate the bright and dark places.

What characterizes a bright point? It is one where two wave crests coincide. This happens when the point is equidistant from the two sources, and also when one source is a whole wavelength farther away than the other. In fact, if L_1 is the distance from the observation point to one source and L_2 is the distance to the other, the two crests will coincide whenever $L_1 = L_2 + m\lambda$, where m is any integer, positive, negative, or zero.

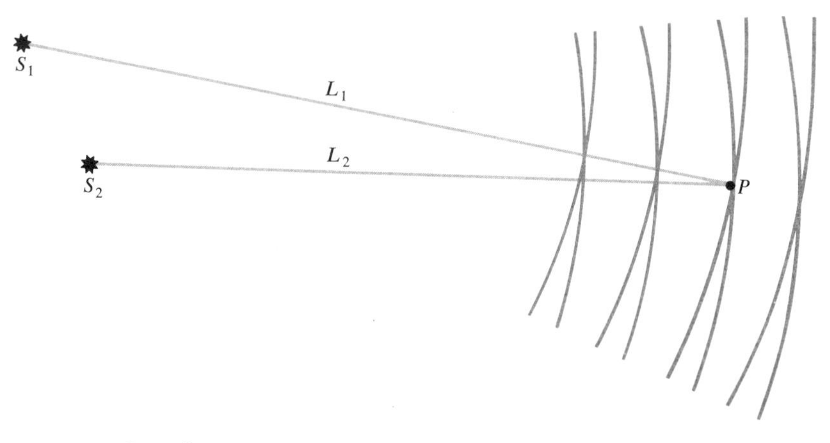

Figure 7.3: Point P is closer to source 2 by $(L_1 - L_2)$.

This takes care of the points in Figure 7.3 where two crests intersect. But the same criterion holds for points where two troughs intersect, or indeed where the waves have the same amplitude. A moment's thought shows that this is as we want it: wherever the waves are identical, they add up to an increased amplitude; we say that they *interfere constructively*. As the waves move out from the sources, the point P experiences successively coincident crests, zeros, troughs, zeros, crests, etc. The interference at the point P is always constructive, although the net amplitude varies with time. Since it is the intensity which interests us in general, we note that the intensity at P is four times the intensity at the same point when either source is turned off.

The intensity at the point Q, however, is always zero. Whatever the amplitude of one wave there, the other always cancels it exactly. Again we can describe the geometric condition that leads to such *destructive interference*. It is the same as before except that one source is an extra half-wavelength farther away. That is, $L_1 = L_2 + m\lambda + \lambda/2$. At a point specified by this condition, the crest of one wave arrives at the same time as the trough of the other, since it has traveled farther at the same speed.

Now these simple pictorial arguments serve to locate the bright and dark points, but we should be able to extract the conditions from our mathematical description of the wave. To do this, we write down the value of E due to each wave:

$$E_1 = E_0 \sin\left[\frac{2\pi}{\lambda}(L_1 - ct)\right]; \qquad E_2 = E_0 \sin\left[\frac{2\pi}{\lambda}(L_2 - ct)\right].$$

The sum of these is

$$E_T = 2E_0 \sin\left[\frac{2\pi}{\lambda}\left(\frac{L_1 + L_2}{2} - ct\right)\right] \cos\left[\frac{2\pi}{\lambda}\left(\frac{L_1 - L_2}{2}\right)\right].$$

The cosine term carries the information about the interference. For instance, it equals 1 when $L_1 = L_2 + m\lambda$, and it equals 0 when $L_1 = L_2 + (m + \frac{1}{2})\lambda$.

7.3 *Average over detection time*

The time dependence in E_T above, is in the sine factor, not in the factor which carries information on the interference. This factor tells us how the wave amplitude changes with time at, say, a bright place. The cosine factor tells us whether or not a place is indeed bright. With visible light we do not expect to perceive the variation with time, since its period is too short for our detectors. Obviously we must detect an average effect such as the average taken over some time T_d, characteristic of our detectors.

Now, if we are detecting a nice slow ocean wave, possibly by bobbing up and down on it, we certainly have a situation in which T_d is much less than the wave's period. In such a case it makes sense to detect the wave amplitude and watch it change with time. But if our detector is a sand bar, which we see being slowly worn away by the same wave, the detection time is too long to give information about when the wave crest and when the trough arrived. All we can say in such a case is that here the disturbance was large and over there it was small. In this case we are *not* averaging the wave's amplitude, but rather the wave energy, which is proportional to the square of the amplitude.

Any of the wave expressions we use give us this result: The average value of E_T is zero unless the time over which we average is small compared with the period. But the average value of $E_T{}^2$ need not be zero.

The average value of a quantity during a time T_d is the sum (or integral) of all values of the quantity, divided by the time T_d. In effect, we add up all positive values, subtract all negative ones, and spread any remainder uniformly over the time interval. Formally, this means

$$\langle E_T \rangle_{av} = \frac{1}{T_d} \int_0^{T_d} E_T(t)\, dt.$$

This integration is worked out in Appendix A, but the formal mathematics is unnecessary here.

We can easily see that

$$\sin\left[\frac{2\pi}{\lambda}\frac{L_1 + L_2}{2} - \frac{2\pi}{\tau}t\right],$$

the time dependent factor in $E_T(t)$, has an average that is zero over one whole cycle. This is indicated in Figure 7.4(a), where the area

Figure 7.4: E_T averaged over three different time spans.

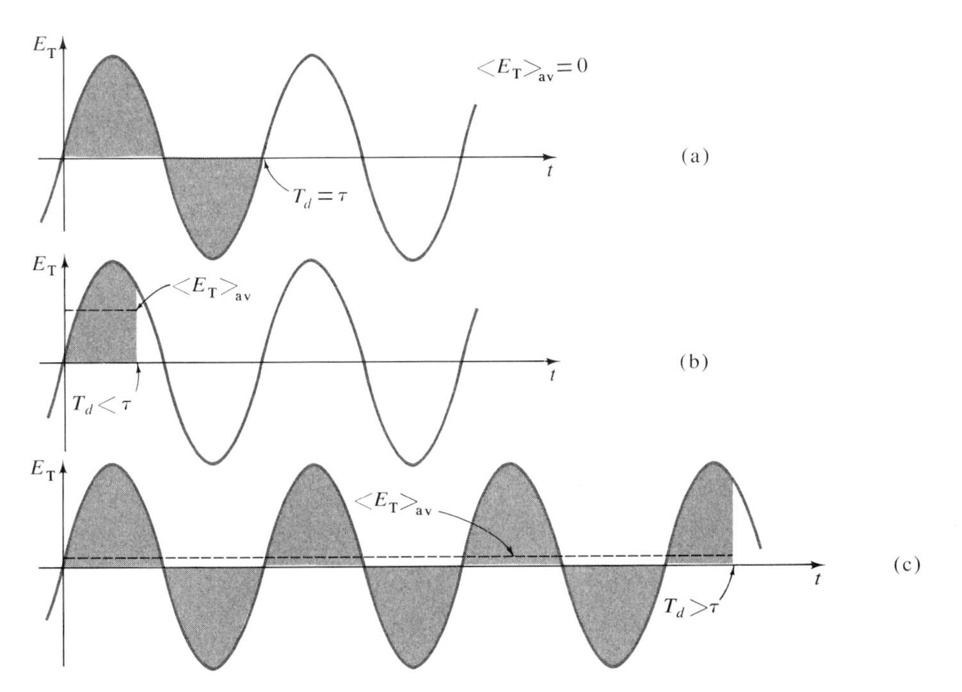

below the axis has been subtracted from that above to give an average of zero. If we average over less than a cycle, the cancellation is not complete, and we get some nonzero average, as in Figure 7.4(b). This is the situation when the time T_d is less than τ. If T_d is much greater than τ, it need not be an even number of cycles, but the net area under the curve is never more than that under a single half-cycle, whereas the time it is spread over may be very large as in Figure 7.4(c). This situation, where $T_d > \tau$, is shown in Figure 7.4(c). The longer the averaging time, the less the average value of E_T. If we measured only E_T, we would never see anything. However, the average value of $E_T^2(t)$ is not zero. $E_T^2(t)$ is never negative, so no cancellations occur. The average over one cycle is just half the peak value, as shown in

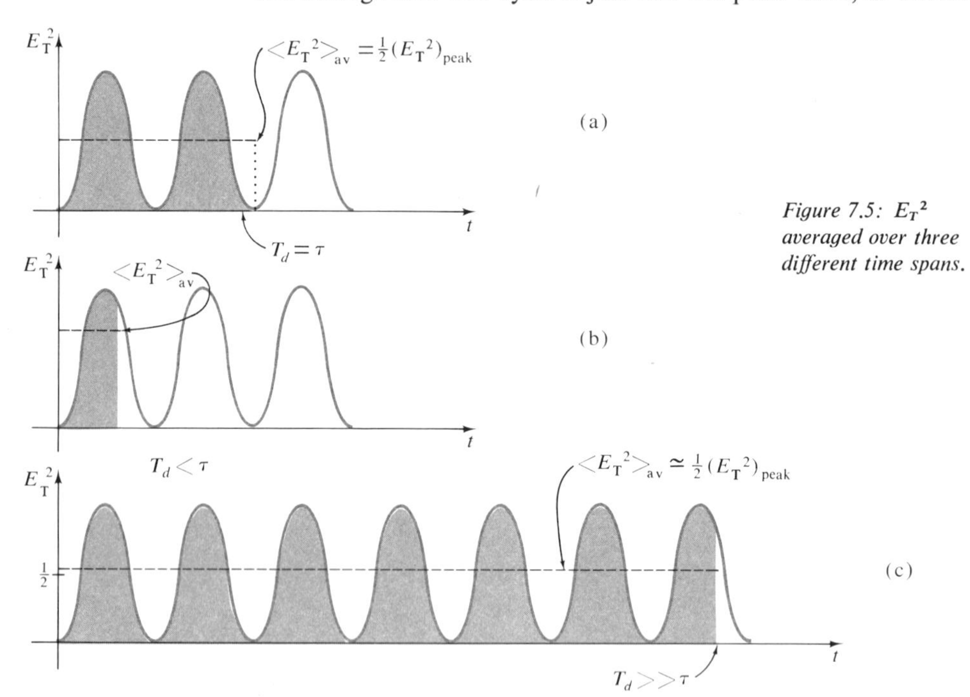

Figure 7.5: E_T^2 averaged over three different time spans.

Figure 7.5(a). A second cycle doubles the area, but also doubles the time span, so the average is the same. When $T_d > \tau$, the average exists and is never as much as the peak value, as is shown in Figure 7.5(b). As T_d becomes much greater than τ, but not an integral number of cycles, the "leftover" area is divided by a very large number, and so contributes little to the average, as shown in Figure

7.5(c). Thus the intensity perceived by our eyes ($T_d \simeq 0.01$ sec) is the same as that detected by a slow photographic film ($T_d \simeq 100$ sec), since for both, $T_d \gg \tau$.

To sum up, if we average over a long detection time (compared with the wave's period), the amplitude's average is zero, while the energy's average is half the maximum possible energy. But if our detection time is short (again in comparison with the period), we perceive the wave amplitude. (7.2)

We can observe this difference not only with water waves, but also with sound and light. For sound, we know that our fingers detect the motion of a loudspeaker diaphragm. When the sound frequency is very low (with respect to our perception time, of course), we can tell whether the diaphragm is in or out at a given instant. But as the frequency is raised, we lose this discrimination, though we can still tell that the diaphragm is in motion. With electromagnetic waves, our eyes can follow the motion of a voltmeter needle up to a few cycles per second. Various electronic devices (fast oscilloscopes, for instance) can follow the wave amplitude up to about 10^{11} cycles per second. Above these frequencies, and in particular for visible light, the detection time is too long, and we must detect energy (or intensity) instead. This is one reason that interference is so important: Here are wave effects of visible light, measurable even with detectors as slow as the human eye or photographic film.

All this tells us what a slow detector perceives. For instance,

Figure 7.6: A film is darkened at points where the interference is constructive.

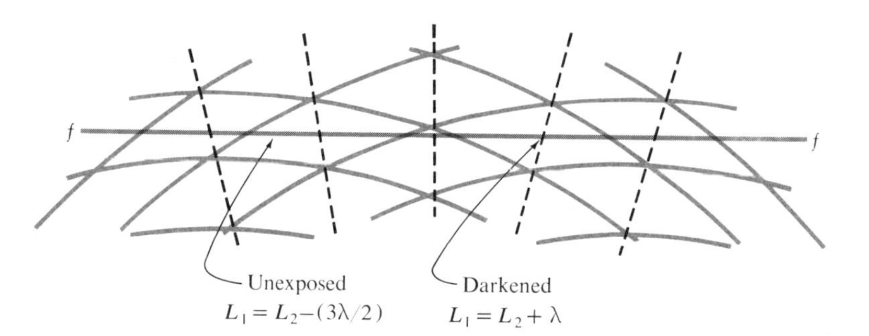

put a piece of film along the line *f–f*, as in Figure 7.6, and it will be blackened most at the points where the interference is constructive, not at all where the interference is destructive, and elsewhere by an amount that we can calculate from our interference equation.

Inspection of the interference equation shows that the intensity variation is a function like

$$\cos^2\left(\frac{\pi}{\lambda}(L_1 - L_2)\right),$$

so that we need only apply some simple geometry to find L_1 and L_2 as a function of position along the line *f–f*. We will, however, defer this until we have considered the question of how we get two 'identical" sources.

7.4 *Coherence*

We have said that source 1 is identical to source 2. This means that they produce waves of the same amplitude, frequency, and *phase*. We have seen how this can be shown if the waves are those of a string—the two sources move up and down in synchronism. We can manage the same thing with electromagnetic waves, of course, by equally charging two metal spheres and waving them up and down, thus generating a wave of as much as 10 Hz, but scarcely of a higher frequency. We can better this, however, by using two radio oscillators that we synchronize in some way (for instance, by comparing them to a common reference signal). This also gets more difficult as the frequency increases, although a device exists (the laser) which synchronizes many atomic emitters at visible light frequencies. An easier system at high frequencies is to use a single oscillator and two antennas. A version of this for sound waves would be a single amplifier and two loudspeakers. At visible wavelengths we might take a single plane wave and let it fall on two pinholes in an otherwise absorbing screen. In all these cases, waves are emitted from two sources that either have the same phase or have a phase difference which is a constant but not a function of time.

Now two atoms will emit waves which have the same frequency and amplitude and which have a phase difference which is constant during the time the atom is emitting. Why don't we consider these to be "identical" sources? An equivalent question is: why don't we observe interference from them? The answer to these questions is again bound up with the length of the detection time. When we averaged intensity, we assumed that $\Delta\phi$ was a constant, and so we took the factor $\cos^2(\Delta\phi/2)$ to be a constant multiplier of the time-dependent function that we averaged. This led to an intensity pattern whose major feature was the $\cos^2$ distribution, which is characteristic of (two source) interference. But the two atoms will have one $\Delta\phi$ all during the emission of one pair of photons and quite a different one during emission of the next pair. So the pertinent value of $\Delta\phi$ *is* a function of time and a random one at that. We would find the same situation with two radio oscillators if their relative phase was selected by setting a dial. If the dial is left alone, then each time the oscillators are turned on they send out waves of the same relative phase so that the bright and dark spots always come at the same places. But if we station an idiot at the dial, who will choose a new value each time, he may choose his values entirely at random. Then the bright spots turn up in different places each time, and we must worry about whether we can get our detecting done before the pattern shifts. If we cannot, then we must expect the interference to "wash out" and be imperceptible, on the average.

We say that two sources with constant relative phase are *coherent*. Let us define the time during which two sources are coherent as the *coherence time* T_c. For instance, the two radio oscillators are coherent as long as the idiot does not change the dial setting, and this is the pertinent coherence time. Similarly, the two atoms are coherent as long as they are emitting the first pair of photons; so the coherence time for an atom is the time it takes to emit a single photon. This time varies among sources, and one of the uses we will make of interference is the measurement of such emission times.

We should formalize these statements by letting $\Delta\phi$ be a random function of time, and averaging E_T^2 over the detection time. Then, if $T_d \ll T_c$, we will see the interference pattern; but if $T_d \gg T_c$ and the $\cos^2$ factor averages to $\frac{1}{2}$, there is no spatial variation of intensity; then

$$I_T = \text{const} \times \langle E_T^2 \rangle_{av} = 2I_0 .$$

This is just the situation we usually see: when we turn on a second light in a room, the intensity at any place just doubles. Similarly, two parts of the same source do not interfere in general, since the atoms in one part are incoherent emitters with respect to those in the other. What this means, of course, is that our detectors are not fast enough to follow the interference patterns which exist momentarily from any source.

A good way to sum up these findings is to remember that when the sources are incoherent, the intensities add; when the sources are coherent, the amplitudes add. In the latter case, of course, the intensity at some points will reach values much greater than the sum of the individual intensities there, with other points receiving no intensity, in such a way that the total energy is conserved. Coherence is a relative concept, and the presence or absence of an interference pattern enables us to measure coherence times if we know the characteristics of the detector. (7.3–7.6)

7.5 *Huygens' principle*

One final aspect of the sources must concern us: we picture the wave as spreading out from the slit or pinhole on which a plane wave has fallen. How can this be, if light travels in straight lines? To answer this, imagine the slit to be filled with some transparent material. The charges in the material scatter in all directions so that if the slit is narrow we will see the waves spreading out from it. Remember that a transparent medium must be of infinite extent in order to scatter only forward. Think now of how an electromagnetic wave is propagated in vacuum: the fields **E** and **B** vary with time at one point and so create new fields in all directions, just as physical scatterers do. The same phasing considerations apply as in a transparent material, and so the wave propagates forward if it is in infinite, empty space.

This way of regarding the light wave as generated by imaginary scatterers in the earlier front was devised by Huygens, and is known as *Huygens' principle*. We will use it extensively and will find from it the precise way in which the wave spreads out from the source, when we study diffraction. For the present, it will be sufficient to work in the Fraunhofer limit, which requires that the waves have traveled far enough from the slits to be essentially plane waves.

EXERCISES

1. Two sources of light of wavelength $\lambda = 0.5\ \mu m$ are separated by a distance $d = 3.62\ \mu m$. If the sources have the same phase, what is the phase difference of the waves, along the line joining them?

2. If the sources in Exercise 1 have different phases and the waves interfere constructively on the line joining them, what is the phase difference of the sources?

3. Estimate the coherence time for ocean waves.

4. Show by Huygens' principle that the common wave fronts (sphere,

plane, cylinder, etc.) preserve their shape as the wave progresses. What happens with noninfinite wave fronts?

5. Show that $I_T = 2I_0$ when $T_d \gg T_c$.

PROBLEMS

7.1 Verify that $I_T = 4I_0 \cos^2 \Delta\phi/2$ for two sources. Show that the energy in a volume that is large compared with a wavelength is unaffected by the interference.

7.2 Two loudspeakers emit a 1000-Hz note. A microphone at a point of constructive interference feeds an oscilloscope. If the oscilloscope sweep rate is 1 cm/msec, the waves are visible and relative phase may be determined by turning off one speaker or the other. What is seen when the sweep is 1 cm/sec?

7.3 Two radio transmitters maintain constant phase, but are turned off each morning, not necessarily at the same time. Under what circumstances are they "coherent"?

7.4 The graph shows the phase $\phi(t)$ of a source that interferes with another source which has constant phase.
(a) Will the eye detect interference?
(b) Will a phototube of 1-nsec response time detect interference?
(c) Are the two sources coherent?

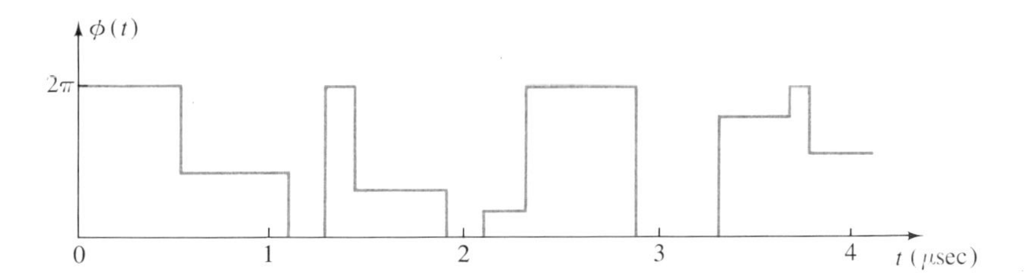

7.5 Between the two sources of Figure 7.1 there is turbulent air. Estimate the magnitude of the time-varying part of the index of refraction necessary to destroy the interference pattern.

7.6 Four sources occur as two pairs. That is, A and B are identical and C and D are identical, but the pairs are unrelated. Find $I_T(z_0)$ when $T_d < T_c$ and when $T_d > T_c$.

7.7 Show that a bright fringe occurs anywhere on the dotted lines of Figure 7.6. That is, when $L_1 = L_2 = m\lambda$, but L_1 need not be an integral number of wavelengths.

7.8 Two loudspeakers are arranged as shown: A is driven by a generator at $\nu_A = 9$ kHz, and with constant phase ϕ_A. B is driven by another generator at frequency ν_B and phase ϕ_B. An observer on the z axis can hear frequencies up to 10 kHz.

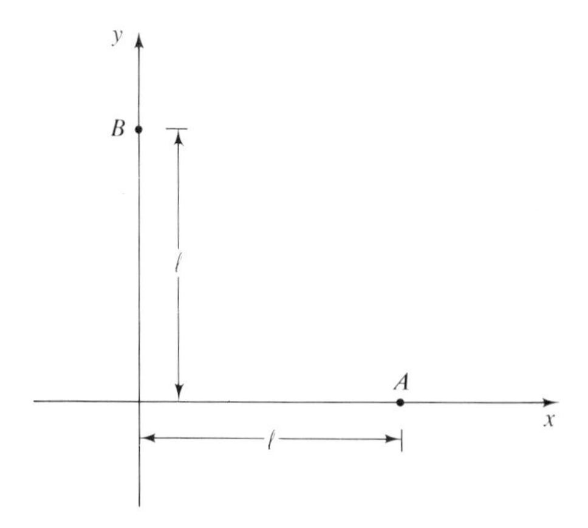

What does he hear if:
(a) $\phi_B = \phi_A$, $\nu_B = 8.95$ kHz?
(b) $\nu_B = \nu_A$, ϕ_B changes to a new value every 10^{-6} sec?
(c) $\nu_B = \nu_A$, and ϕ_B changes between ϕ_A and $\phi_A + \pi$ every 10^{-2} sec?

8

Interference from two sources

This chapter introduces a specific interference problem which is often encountered: two close sources, observed in the Fraunhofer limit. Many of the characteristics of general interference patterns are exhibited by this simple one. In this chapter the problems represent most of the standard two-source problems of optics.

8.1 *Identical sources*

We consider the interference of plane waves from two sources. Figure 8.1 shows the geometry. The two sources are separated by a distance d. The observation point, P, is a distance L_1 from one source

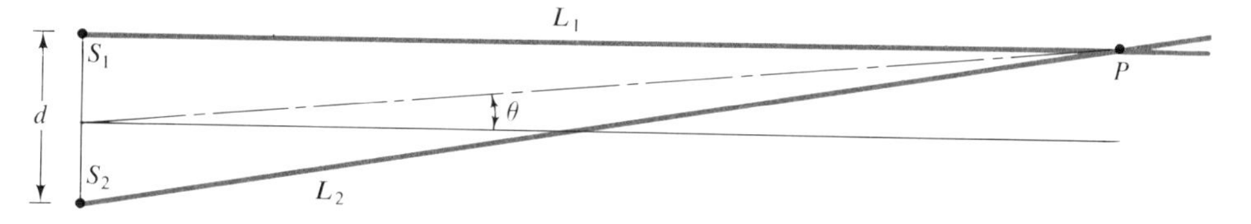

Figure 8.1: The standard geometry for "two-slit" problems.

and L_2 from the other. A line from P to the midpoint between the sources makes an angle θ with the normal to the line joining the sources. Now the Fraunhofer condition requires that the waves be plane waves at the point P. This is equivalent to saying that $\lambda \ll L_1$ or L_2.

For simplicity we will also choose d small compared with these distances, L_1 and L_2 so that we can regard L_1 and L_2 as parallel and also as making the same angle θ with the normal to the line joining the sources. If P were on the normal ($\theta = 0$), L_1 would be equal to L_2. But if θ is not zero, we can find $L_2 - L_1$ from Figure 8.2.

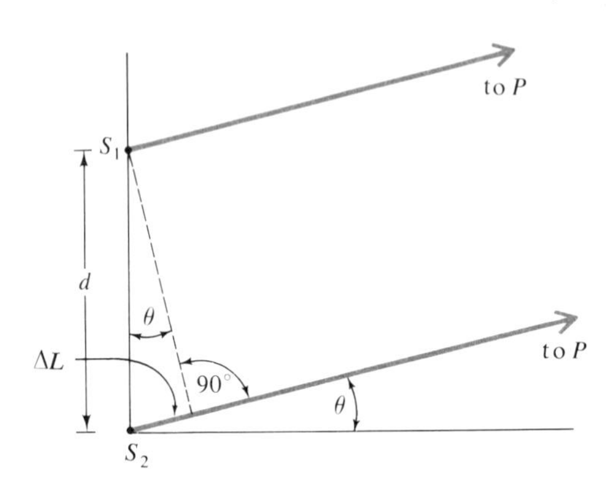

Figure 8.2: Detail of the standard geometry.

$$L_2 - L_1 = \Delta L = d \sin \theta.$$

Then, adding the two waves at P,

$$E_T(P) = 2E_0 \sin\left\{\frac{2\pi}{\lambda}\frac{L_1 + L_2}{2} - 2\pi\nu t\right\} \cos\left\{\frac{2\pi}{\lambda}\frac{\Delta L}{2}\right\}$$

from which we see that

$$\langle E_T(P)\rangle_{av} = 0, \qquad \langle E_T{}^2(P)\rangle_{av} = 4E_0{}^2 \cdot \frac{1}{2} \cos^2\left(\frac{\pi d \sin \theta}{\lambda}\right).$$

Thus the intensity at the point P is $I = I_0 \cdot 4 \cos^2(\pi d \sin \theta/\lambda)$, where I_0 is the intensity of one of the sources alone, at the point P. A particularly useful part of this is that we can now use the interference pattern to measure optical wavelengths. For instance, the bright regions occur when $\sin \theta = m\lambda/d$, and the dark ones when $\sin \theta = (m + \frac{1}{2})\lambda/d$. So the successive dark (or bright) regions are separated by $\Delta(\sin \theta) = \lambda/d$. Thus, if we know d, a measurememt of the angles of two dark (or bright) regions is sufficient to find the wavelength. Usually the sources are slits, and the bright and dark regions are lines, called "fringes".

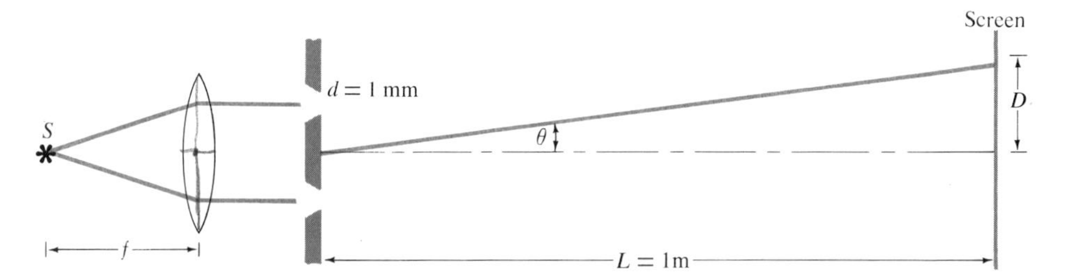

Figure 8.3: Two-slit interference pattern on a distant screen.

Consider the experimental apparatus shown in Figure 8.3: How far apart are the fringes? The zeroth bright fringe is at the center point, $\theta = 0$. The next one, $m = 1$, is at $\sin \theta = 1 \cdot \lambda/d = 5 \times 10^{-4}$ rad. This is in the region where θ is small enough to say $\sin \theta = \tan \theta$. We measure $\tan \theta$ directly: $\tan \theta = D/L$, where D is the distance to the first fringe, which we wish to calculate. Thus, $D = L \times 5 \times 10^{-4} = 0.5$ mm. This fringe spacing is readily discernible with the naked eye. Of course, if θ is large, we will want a curved screen, and the small angle approximation will no longer be appropriate. (8.1–8.3)

Often we wish to work in the Fraunhofer limit without placing the observation point so far from the sources. The plane waves we want are then brought to form an image on a screen at the focal plane of the lens as shown in Figure 8.4. If plane waves illuminate the

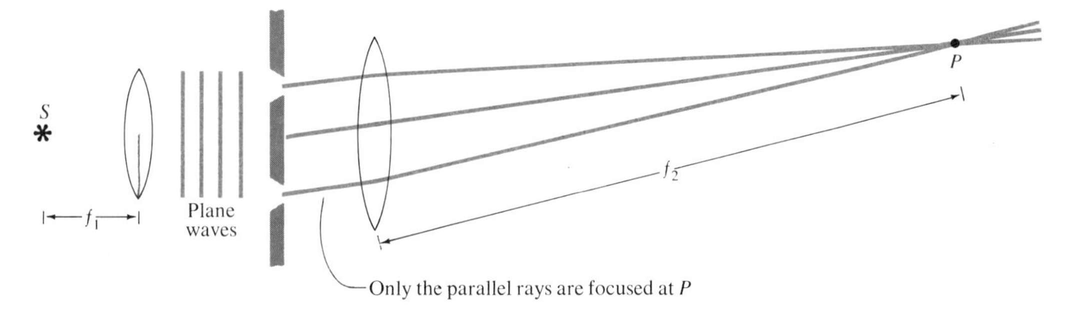

Figure 8.4: With a lens, the Fraunhofer condition is satisfied for a close screen.

two slits, the fringes are images of a point source. More commonly, the slits are illuminated by cylindrical waves from a line source at the focus of a lens. Then the images are lines. If we follow the undeviated ray through the center of the lens, the geometry is the same as before. This is treated in a problem at the end of this chapter.

The two-source problem is subject to practically infinite variation. The coherent sources, for instance, may be formed in many ways, some of which are illustrated in the problems. Others are described in the books listed in the Bibliography. We may also modify the assumption of identity for each of the parameters of the two waves: phase, velocity, frequency, and amplitude. Instead of worrying about details of the sources, here we will consider two sources which are not in phase, a two-source problem where the light paths are through materials of different refractive index, sources of different frequency, and sources of different amplitude.

8.2 *Sources differing in phase*

The two sources may differ in phase by a constant amount and still be coherent. This may occur because of some phase shift built into the sources, or simply because the incident plane wave does not fall normally on the two slits. Then we must add a term $\Delta\phi_{sc}/2$ to the argument of the interference factor:

$$I = I_0 \cos^2\left(\frac{\pi \, \Delta L}{\lambda} + \frac{\Delta\phi_{sc}}{2}\right).$$

This simply shifts the fringes sideways. We can, for instance, choose the relative phase of two loudspeakers to be π. Then the "bright fringes" shift to the former position of the dark ones. (8.4–8.5)

8.3 *Paths differing in index*

Another variation that we can make is to introduce refracting media between the slits and the observation point. This, of course, has no effect on the sources, and their identity or lack thereof. In fact, only the phase velocity c is changed in the description of the waves: the effect of the index of refraction is to substitute c/n for c. We can think of this as changing λ to λ/n. Let us therefore fill the two paths with materials of index n_1 and n_2. The "central" bright fringe, which used to occur at the position $L_1 = L_2$, now comes at the place where $L_1 n_1 = L_2 n_2$, since crests which started together arrive there together. Notice that the important quantity is really the *time* a wave takes to travel from source to observation point.* Or we can think of the two paths as containing the same number of wavelengths: $L_1 = N(\lambda/n_1)$, and $L_2 = N(\lambda/n_2)$. This latter is a convenient pictorial way of understanding the problem. (8.6–8.9)

* This observation forms the basis for a very general statement known as Fermat's principle. This has important ramifications, and is treated very well in Feynman's book (see Bibliography).

8.4 *Sources differing in frequency*

What if the two sources differ in frequency? This is a more basic variation of the simple problem, but is solvable by considerations we have already developed. If the difference in frequency is $\Delta\nu$, then our intensity equation becomes:

$$I = I_0 4 \cos^2\left(\frac{\pi \Delta L}{L} + \frac{\Delta\phi}{2} + \pi \Delta\nu t\right).$$

This presents us with the same sort of situation encountered when considering coherence: we must average this time-dependent interference pattern over our detection time. As before, we will lose the interference effects if the time variation is too fast with respect to the detection time. Here, that means we will see interference only if $T_d < 1/\Delta\nu$. If we do see a pattern, it will have *beats* in time, a situation physically realizable only for very special sources like lasers. (8.10–8.11)

8.5 *White light*

The preceding discussion leads us to ask how it is possible for us to see interference effects with white light. The primary difference, of course, is that the white light comes through *both* slits. The red light forms an interference pattern with a given spacing, and the blue light (incoherent to the red in the sense just discussed) forms another pattern with a different spacing. When the patterns coincide (dark fringes falling on dark fringes), they are both visible. But if they do

Figure 8.5: Overlapping of fringes formed by light of different colors.

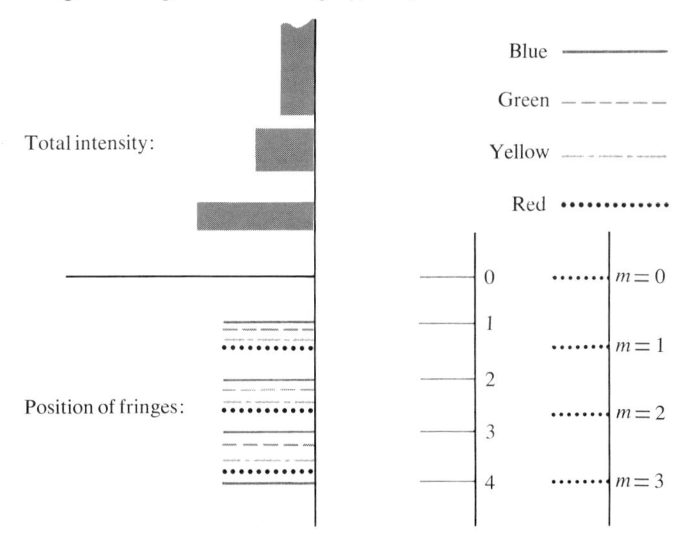

not coincide, a colored blur results. Figure 8.5 shows how the visibility decreases for white-light fringes as the observation point gets far from the equal path-length condition.

In the upper half of the fringe patterns of Figure 8.5, we see the total intensity, regardless of color, as black-and-white film would show it. In the lower half, the individual fringes are shown, out to the third red one and the fourth blue one, which overlap. (8.12)

8.6 *Phasors* Before going on to three and more sources, we ought to note the way in which phasors may be used to describe the interference pattern, since this approach will be most useful with the multiple sources. For each slit we draw a phasor whose length is proportional to the square root of the intensity from that source. The angle made by two phasors is the difference in the arguments of the two waves: $\gamma = 2\pi\, \Delta L/\lambda + \Delta\phi$. The argument of the $\cos^2$ factor in the intensity equation is just $\gamma/2$. Figure 8.6 illustrates this for some simple cases.

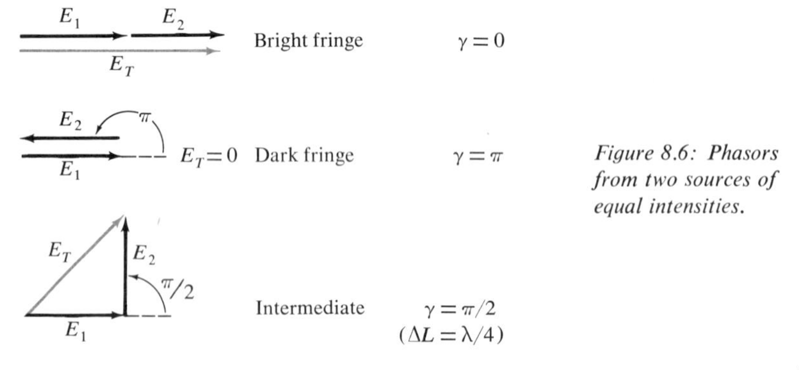

Figure 8.6: Phasors from two sources of equal intensities.

If the slits are of unequal width or one is somewhat opaque, the intensities will be different. Then the phasors are shown as in Figure

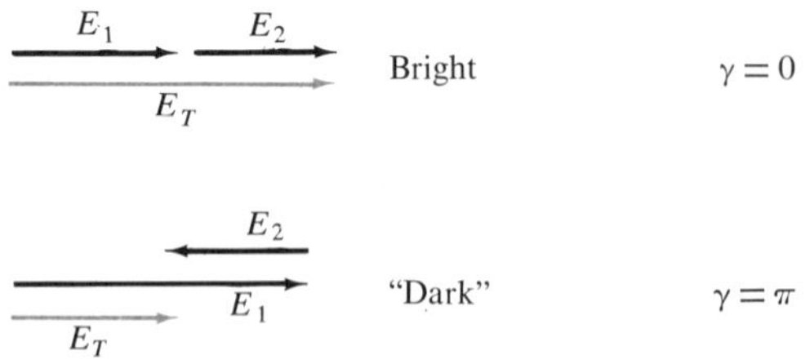

E_1 E_2 E_T Bright $\gamma = 0$

E_2 E_1 E_T "Dark" $\gamma = \pi$

Figure 8.7: Phasors from two sources of different intensities.

8.7. Notice that the minimum is still at $\gamma = \pi$, but that $E_T \neq 0$ there, since the second wave can cancel only part of the first.

We have seen how two sources can produce a pattern of varying intensity under certain conditions. We require that they be coherent with respect to each other; that is, that they maintain a phase relationship which is constant for a time long enough for our detectors to register the pattern. We say that the coherence time T_c must be longer than the detection time T_d. We have studied the pattern formed in the Fraunhofer limit, where the observation point is so far from the sources that the waves may be thought of as plane. Having found the pattern, we have used it to study various differences between the sources or the paths the light can follow.

EXERCISES

1. If $L_1 = 10d$, what error is introduced by the assumption that the two rays are parallel, for $\theta = 0.01$? For $\theta = 0.1$? For $\theta = 1$?

2. In the standard geometry, fringes are 10 μm apart and 3 m from the slits. Find λ if $d = 1$ mm.

3. White light is viewed through two slits 0.5 mm apart, held close to the eye. Estimate the fringe spacing on the retina.

4. Two slits are illuminated by plane waves from a source at an angle α form their normal. Find $I(\theta)$.

5. Fringes from two slits are observed, using light of wavelengths $\lambda_1 = 0.40$ μm and $\lambda_2 = 0.45$ μm. At what m values do the fringes of the two colors coincide?

6. Two rocks at 1 km off shore obstruct incoming ocean waves. At the beach the waves show interference fringes 20 m apart. The wavelength is 1.5 m. How far apart are the rocks?

PROBLEMS

8.1 A radio telescope sits on a cliff, 100 m above the ocean, and detects radiation from a star at an angle α above the eastern horizon. Plot the intensity variation at the receiver, as the earth turns. Since the ocean acts like a mirror, this is a Lloyd's mirror interferometer.

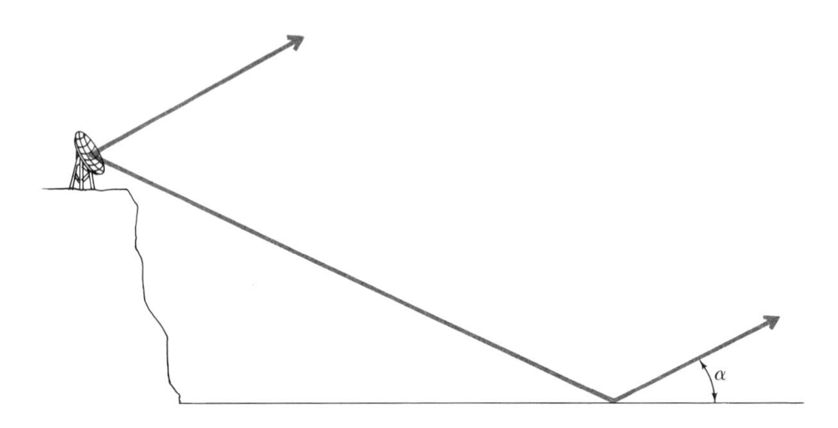

8.2 A laser beam illuminates two prisms, as shown. The beam consists of monochromatic plane waves and has a circular cross section 10 cm in

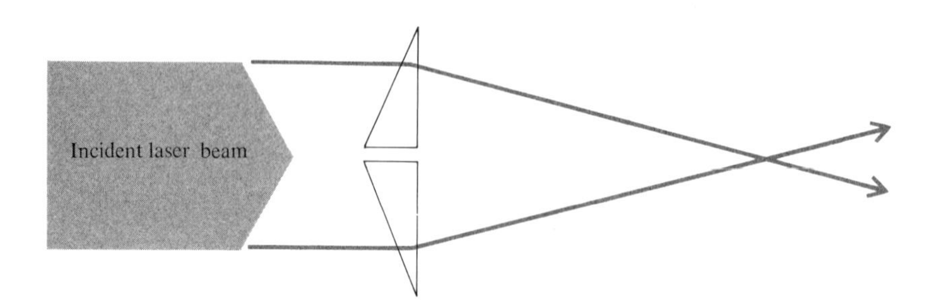

diameter. The index of refraction of each prism is 1.5, and their angle is 1 degree. Sketch the interference pattern 5 m away, and describe the fringes mathematically. (Fresnel biprism.)

8.3 A laser beam illuminates two mirrors, as shown. If we assume that the beam consists of strictly monochromatic plane waves and has a circular cross section of 10-cm diameter, what interference pattern do we see at a distance of 5 m from the mirrors? The angle between the mirrors is 0.01

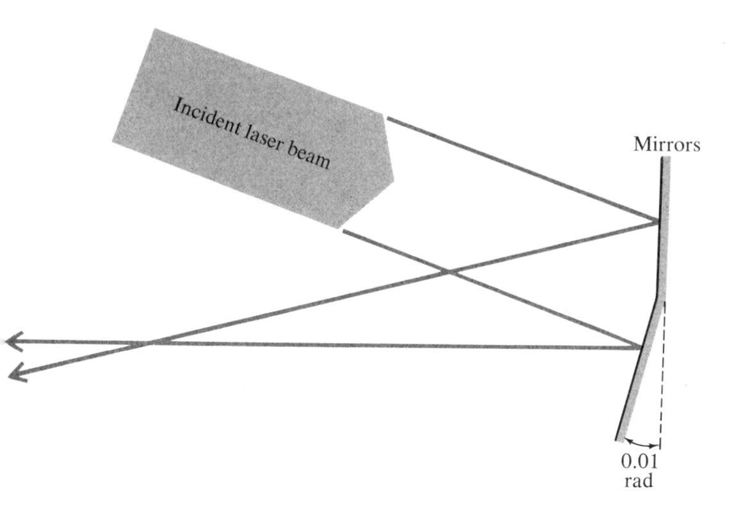

rad (about $\frac{1}{2}$ degree). Sketch the pattern and give a mathematical description of the fringes, where they occur. (Fresnel double mirror.)

8.4 The figure shows a double-slit apparatus illuminated by a monochromatic source of finite size. Describe the interference patterns.

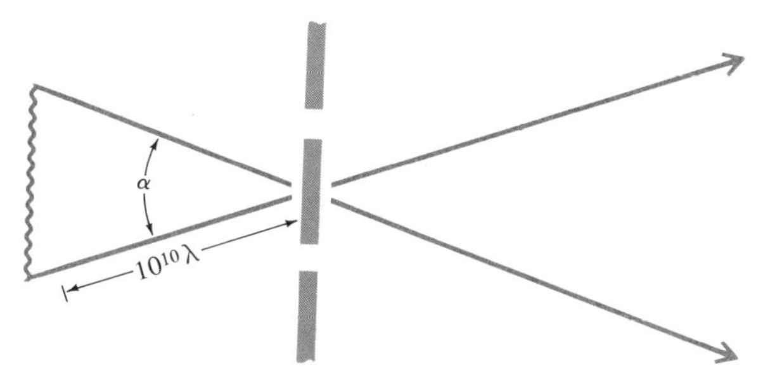

8.5 The stars Betelguese and Rigel are equally bright, and are 17.1 degrees apart. Their light falls on the apparatus shown, where F is a filter passing only the wavelength λ, and S is a pair of slits with separation d. Find the total intensity as a function of the angle θ.

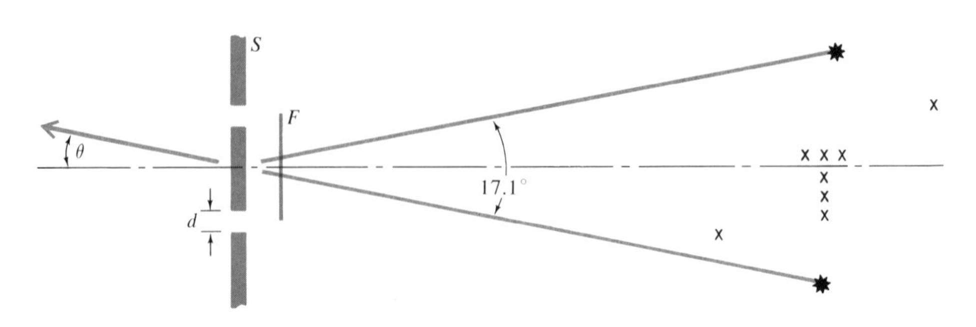

8.6 The figure shows an arrangement for measuring the index of refraction of gases. The two slits are illuminated by monochromatic plane waves from

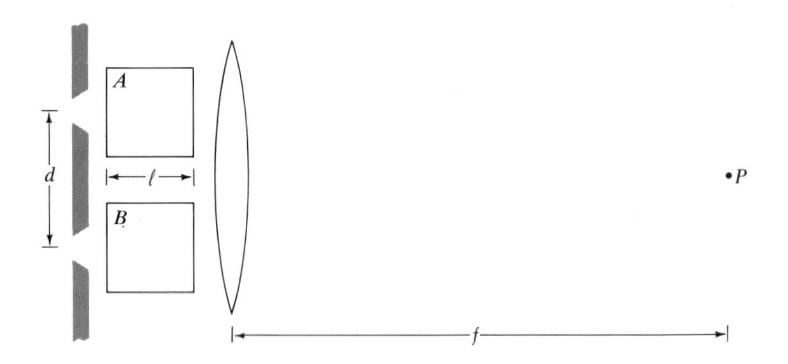

the left. To the right of the slits are two identical glass containers A and B, each of inside thickness ℓ. With both containers evacuated, a bright fringe appears at P on a screen opposite the center of the slits. A gas is then admitted to A, resulting in a shift of 20 fringes as observed at P for light of wavelength λ.

(a) Which way did the fringes move?
(b) What is the index of refraction of the gas?

8.7 In the apparatus of Problem 8.7, a flat piece of glass (index 1.5) is substituted for cell B. By rotating the glass through an angle α, the fringe shift due to the gas is removed. How thick must the glass be?

8.8 A double-slit experiment is performed, with the modification that following slit A is a $\lambda/2$ plate with fast axis along the slit, and following slit B is a $\lambda/2$ plate with fast axis perpendicular to the slit. The light is unpolarized. What is the position of the dark fringes?

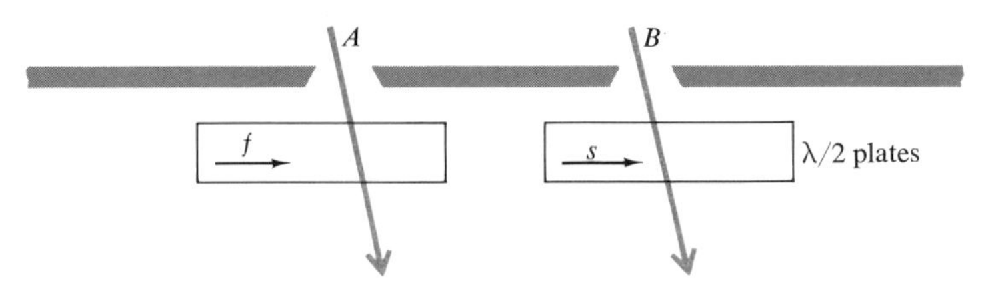

8.9 Quarter-wave plates are substituted for the half-wave plates of Problem 8.9. What pattern is observed?

8.10 One of the mirrors in Problem 8.3 is unstable and rotates slowly at angular velocity $\omega = d\alpha/dt$. What happens to the fringe pattern? How large can ω be if we are to detect the interference on common photographic film?

8.11 Two flutists sit 8 m apart and both play middle C.

(a) Where should a music-hater sit to avoid hearing them?

(b) In the heat of the music, one flute expands and plays B. How fast do the fringes move across the audience?

8.12 A source for a double-slit experiment emits light of wavelengths λ_1 and λ_2 with equal intensities I_0. The geometry is the usual one, with a slit separation of d. Two filters are available: F_1 lets through only λ_1, and F_2 lets through only λ_2.

(a) If both slits are covered by F_1, at what angle is the ninth *bright* fringe?

(b) If both slits are covered by F_2, at what angle is the ninth *dark* fringe?

(c) If slit A is covered by F_1 and slit B by F_2, what is $I_T(\theta)$?

(d) If no filters are used, find the first angle greater than $\theta = 0$, where the total intensity has its maximum value. Use $\lambda_1 = 5000$ Å, $\lambda_2 = 4000$ Å, and $d = 1$ mm.

8.13 Two identical sources are far apart, but still a large number of wavelengths from an observer. Thus the Fraunhofer condition is satisfied, but the approximation of parallel rays from the sources to the observer is not. Find the fringe spacing near the observation point when: (a) observer and source are nearly collinear, (b) observer and sources are at the vertices of an equilateral triangle, and (c) observer and sources lie on the circumference of a circle with the sources on a diameter.

8.14 One of the mirrors in Problem 8.3 is allowed to receed at a speed v, which is small compared with the velocity of light. What happens to the fringe pattern? How large can v be if we are to perceive the interference with our eye?

9 *Interference from many sources*

We proceed now to the problem of interference of more than two waves. As with the case of two sources, there are many variations possible on the basic problem. However, we confine ourselves here to the simplest version: All sources are identical in every respect except their distance from the observation point. We will use the phasor approach to calculate the position of the bright and dark fringes, and later the intensity variation.

9.1 *Three slits*

Consider first the three slits shown in Figure 9.1: when $L_1 = L_2 = L_3$, the phasors all line up and we have a bright fringe. With two slits, we had a dark fringe when $L_1 = L_2 + \lambda/2$. Then the phasors lined up in opposition.

$$\overset{E_2}{\longleftarrow}\quad \underset{E_1}{\longrightarrow}$$

With three slits, such a path difference would indeed let E_1 cancel E_2, but would leave E_3. The phasors look like this:

$$\rightleftarrows\!\!\rightarrow \quad \text{with } E_T = E_3 = \frac{E_{T\max}}{3}.$$

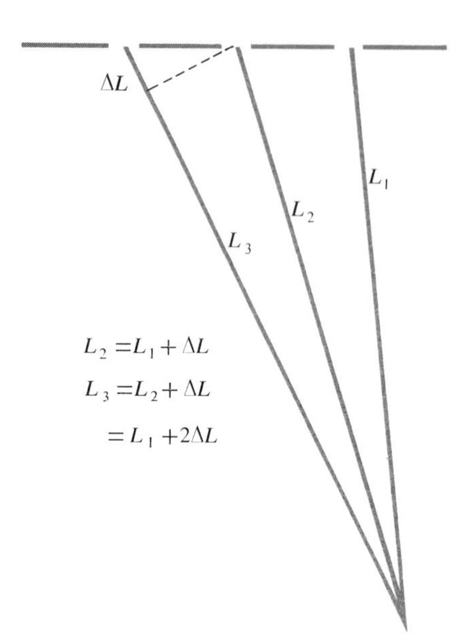

Figure 9.1: Three identical sources.

Consequently, $I_T = I_{T\max}/9$. Then where is the dark fringe? This must come where the resultant of the three phasors is zero: The angle γ, between adjacent phasors, must be $2\pi/3$. The origin of γ is the path difference:

$$E_1 = E_0 \sin\left[\frac{2\pi L_1}{\lambda} + 2\pi\nu t\right],$$

$$E_2 = E_0 \sin\left[\frac{2\pi L_2}{\lambda} + 2\pi\nu t\right],$$

and

$$E_3 = E_0 \sin\left[\frac{2\pi L_3}{\lambda} + 2\pi\nu t\right].$$

At the observation point, the arguments differ by

$$\gamma = \frac{2\pi(L_1 - L_2)}{\lambda} = \frac{2\pi(L_2 - L_3)}{\lambda}.$$

This means that the dark fringe occurs when $\gamma = 2\pi/3$ or when $\Delta L = \lambda/3$. A second dark fringe is found at $\gamma = 4\pi/3$.

Having located the bright and dark fringes, we can find some other convenient points, such as $\gamma = \pi/2$, as in Figure 9.2a. Only a few values are needed to make a sketch of the intensity curve, as shown in Figure 9.2b. (9.1–9.4)

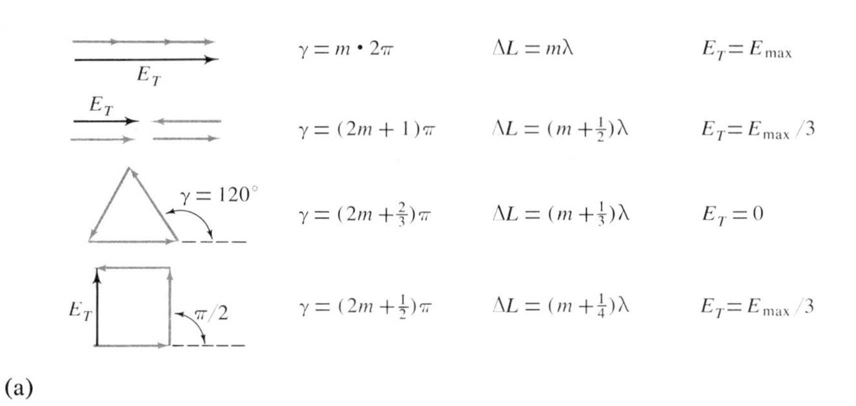

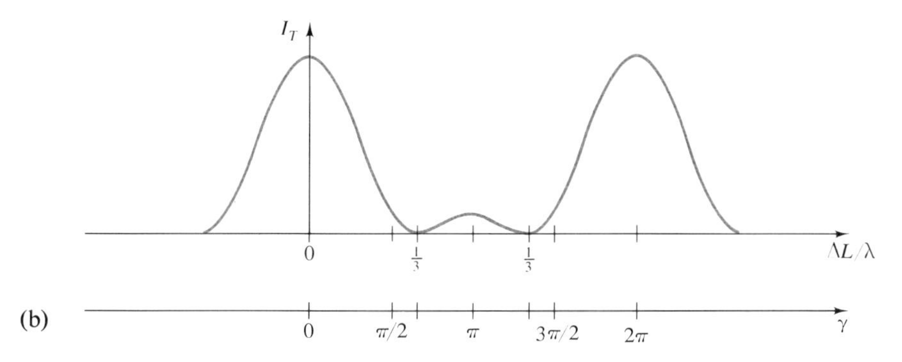

Figure 9.2: Total intensity at the observation point for three slits.

Notice the appearance of a secondary maximum. As we go to systems of more slits, there will be more secondary maxima between the primary ones.

The case of four slits is given as a problem at the end of this chapter. Again we see secondary maxima, this time 16 times less intense than the primary ones. (9.5)

9.2 *Grating* The general case of N slits is called a grating. We can treat it in the same way we have treated the special cases with $N = 2$ and 3. Thus the distance from the observation point to one slit differs from the distance to an adjacent slit by ΔL and we can write $\gamma = 2\pi\,\Delta L/\lambda$. For the usual geometry, $\Delta L = d\sin\theta$, with d the distance between adjacent slits. As usual, the equidistant point, $\Delta L = 0$, is a bright fringe, with N phasors all lined up. This situation is then repeated when each phasor has turned through 360 degrees:

$\rightarrow\rightarrow\rightarrow\rightarrow\rightarrow\rightarrow\rightarrow\rightarrow\rightarrow\rightarrow$ (phasors E_i, E_i) $\quad \gamma = 0, \quad \Delta L = 0, \quad E_T = NE_i, \quad I = N^2 I_i,$

$\rightarrow\rightarrow\rightarrow\rightarrow\rightarrow\rightarrow\rightarrow\rightarrow\rightarrow\rightarrow \quad \gamma = 2\pi m, \quad \Delta L = m\lambda, \quad I = N^2 I_i.$

These points, of maximum intensity, are called *principal maxima.* As with the three-slit case, the first zero is at the point where $\gamma = 2\pi/N$; therefore each of the N phasors has turned $(1/N)$th of 360 degrees. The Nth phasor completes a closed figure and the resultant is zero. This occurs $N - 1$ times between principal maxima, as

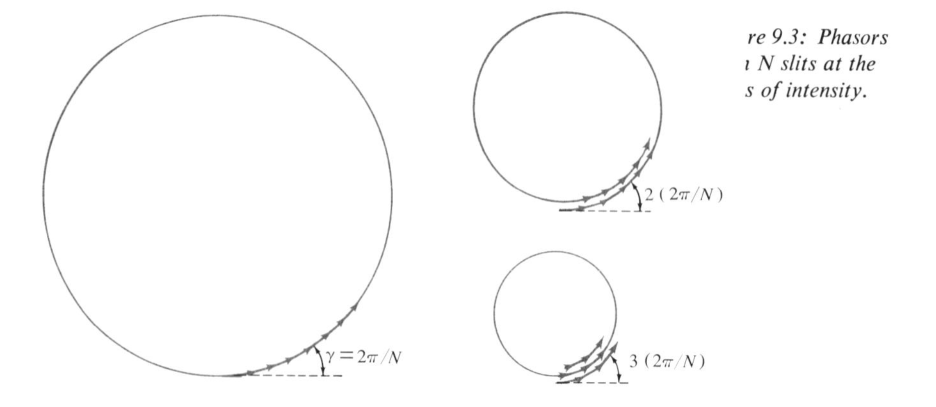

re 9.3: Phasors ı N slits at the s of intensity.

indicated in Figure 9.3. Thus the zeros are equally spaced. Every Nth zero is replaced by a principal maximum, whenever m' is an even multiple of N.

$$\gamma_{\text{zero}} = m' \cdot \frac{2\pi}{N}, \qquad m' \neq mN.$$

Now, somewhere between zeros, there must be other maxima in the intensity curve, since E_T is not zero all the time. These secondary maxima come about halfway between the zeros. They diminish in size rather rapidly so that we find a pattern like that of Figure 9.4.

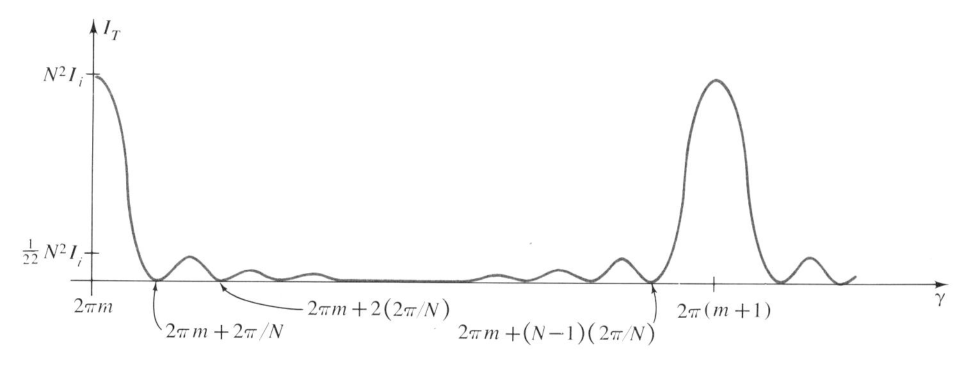

Figure 9.4: Total intensity at the observation point for N slits.

The main features of the intensity curve are already given by our qualitative discussion: the intensity is small except at a few values of γ, where it becomes very large over a narrow range.

9.3 *Line width*

We will want to know how narrow is this range of γ values. That is, we want to find the "half-width"* of a principal maximum, the distance along the curve from the peak to the first zero. We can see from Figure 9.4 that this distance is

$$\begin{aligned}(\Delta\gamma)_{1/2} &= \gamma_{\text{zero}} - \gamma_{\text{peak}} \\ &= (mN+1)\cdot\frac{2\pi}{N} - mN\cdot\frac{2\pi}{N} \\ &= \frac{2\pi}{N}.\end{aligned}$$

So the half-width is

$$(\Delta\gamma)_{1/2} = \frac{2\pi}{N} \qquad \text{or} \qquad (\Delta L)_{1/2} = \frac{\lambda}{N}.$$

As we might expect, the principal maxima get narrower as N increases. However, this means "narrow relative to the separation

* Other uses of the term "half-width" refer to the width of the curve between points of half maximum amplitude or intensity. These "half-widths" are respectively 0.9 and 1.2 times as wide as ours.

between principal maxima". To see this, plot $I(\sin\theta)$ and observe the half-widths $\Delta(\sin\theta)_{1/2} = \lambda/Nd$. Since $(N-1)d$ is the distance between first and last slits, we get the situation shown in Figure 9.5. Notice that if we use just the first and last (Nth) slits of the grating, we get half-widths nearly as narrow as with all N. But the intensity is smaller and the resolution (the *relative* narrowness of the maxima) is much less. (9.6)

Figure 9.5: $I(\theta)$ for N slits, illustrating relative spectral linewidths due to slit spacing and number.

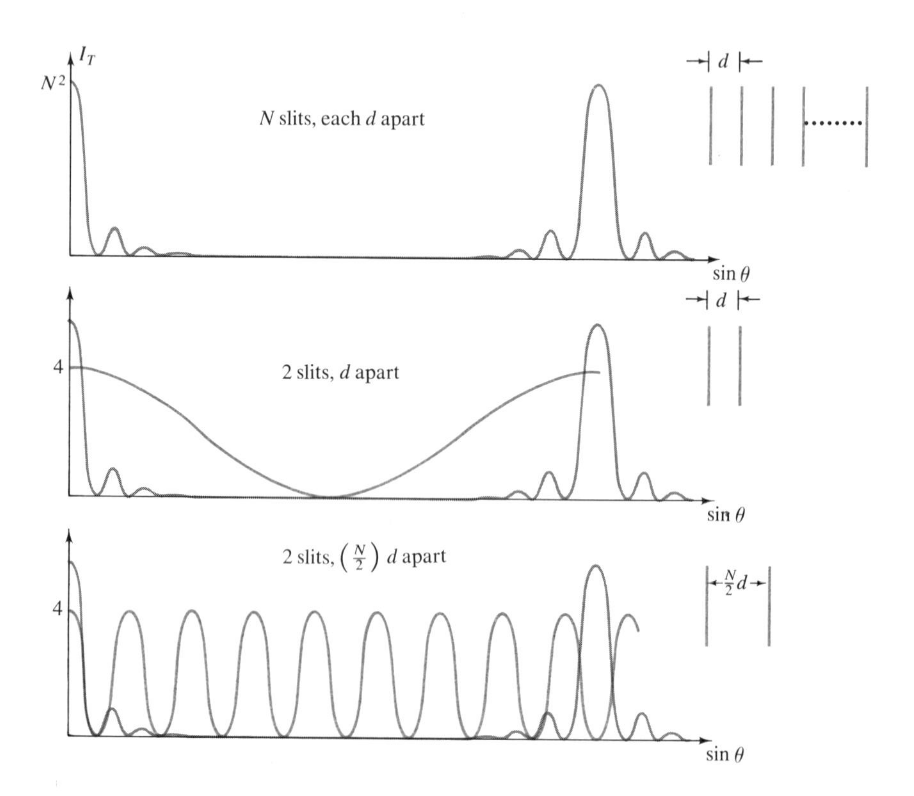

9.4 Grating equation

We can use our phasors quantitatively to find the exact form of the I versus γ curve (or I versus θ): in Figure 9.6, N phasors, each of length E_i, add up to the resultant E_T. Since each phasor turns through γ relative to its predecessor, the total angle between first and last is γN. We can then give two expressions for H, equate them, and find E_T.

From Figure 9.6(a),

$$\frac{1}{2}E_T = H \sin\frac{N\gamma}{2}$$

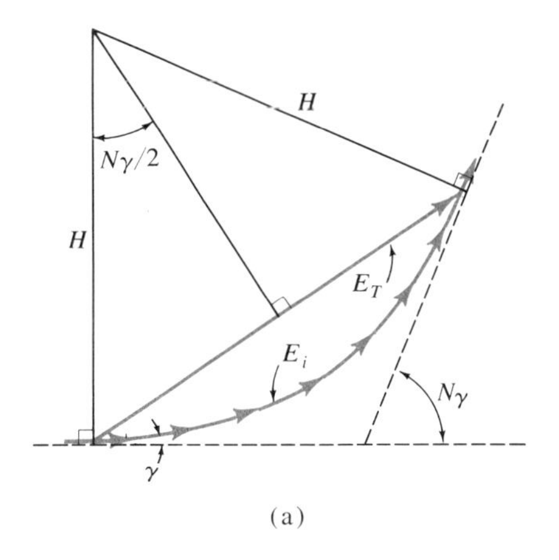

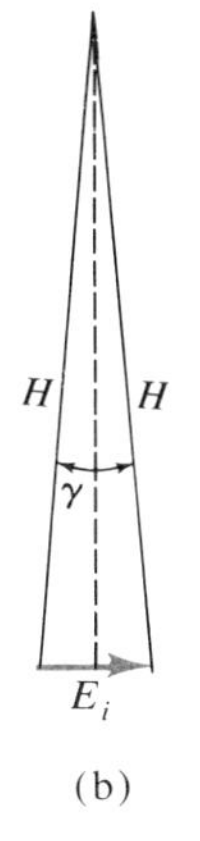

Figure 9.6: The geometry for finding E_T from N equal phasors, each differing in phase angle by γ from its predecessor.

From Figure 9.6(b),

$$\frac{1}{2}E_i = H \sin\frac{\gamma}{2}.$$

Eliminating H:

$$E_T = E_i \frac{\sin(N\gamma/2)}{\sin(\gamma/2)}.$$

Thus the general expression is

$$I(\gamma) = I_i \frac{\sin^2(N\gamma/2)}{\sin^2(\gamma/2)}.$$

It is left to the student to verify the features of the $N = 2$, 3, and 4 curves from this, as well as the N slit curve we had previously sketched. (9.7)

9.5 *Wavelength resolution*

Let us go back to the equation which specifies the position of the principal maxima: $\sin\theta = m\lambda/d$. The principal maximum for red light occurs at a larger angle than that for blue light. So, if we illuminate the grating with white light as in Figure 9.7 each principal maximum will be spread over the range of angles between these values.

When atoms of tenuous gases emit light, they generally do so at a

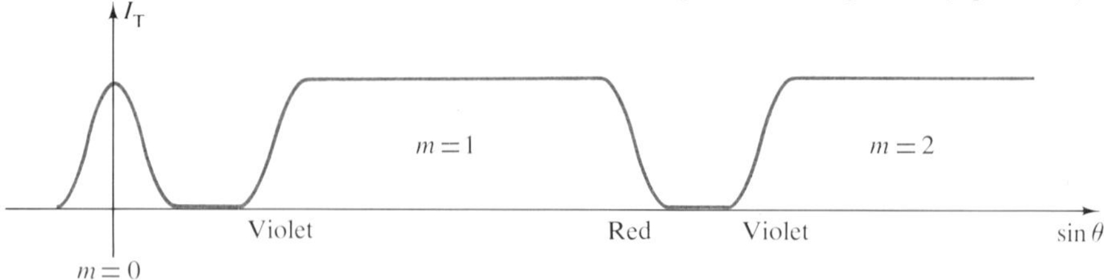

Figure 9.7: Wavelength separation by a grating.

small number of wavelengths. Gratings are used as spectrometers, to sort out these wavelenths, and the principal maximum for a given wavelength is called a spectral "line". Naturally, we want such lines to be narrow enough so that we can tell when there are two of them and when there is only one. If this condition is met, we say that the lines are *resolved*. Representative cases are illustrated in Figure 9.8.

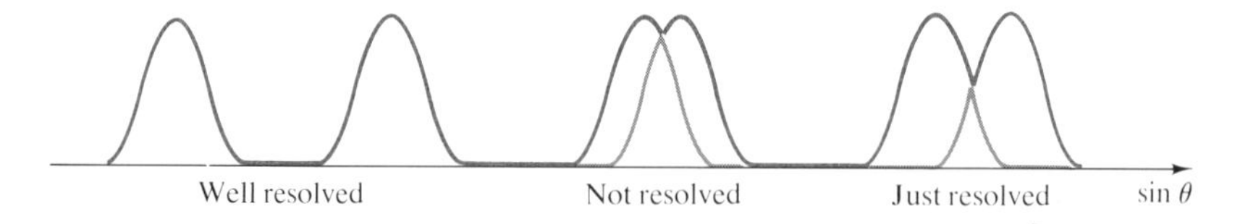

Figure 9.8: Resolution of spectral lines.

If two lines are separated in wavelength by $\lambda_1 - \lambda_2 = \Delta\lambda$, the corresponding separation in angle is $\Delta(\sin\theta)_\lambda = m\,\Delta\lambda/d$. As a somewhat arbitrary definition of resolution, we will say that a line is resolved from its neighbor if $\Delta(\sin\theta)_\lambda$ is greater than the half-width. This means that if two lines are far enough apart so that the peak of one is no closer than the first zero of the next, we may expect to be able to tell that there are two of them. Thus the smallest separation detectable with a given grating is that for which $\Delta(\sin\theta)_\lambda \geq \Delta(\sin\theta)_{1/2}$, which means that $m\,\Delta\lambda/d \geq \bar{\lambda}/Nd$, where $\bar{\lambda}$ is an average wavelength, $(\lambda_1 + \lambda_2)/2$.

The integer m, which is called the *order* of the spectrum, is usually restricted to 2 or 3 by diffraction effects (which we have yet to study) and by the fact that $\sin\theta$ must always be less than 1. A further complication is that for larger values, the spectra in the various orders overlap so that the red in third order may fall at the same angle as the violet in fourth.(9.8–9.10)

9.6 *Broadening*

Before we leave the subject of gratings and spectral lines, we should note that the width of lines as they are actually observed may depend on a number of factors other than the resolution of the grating. Another *instrumental* effect is the width of the entrance slit of the spectrometer; that is, the slit which produces the plane wave that falls on the grating in the first place. As shown in Figure 9.9, the

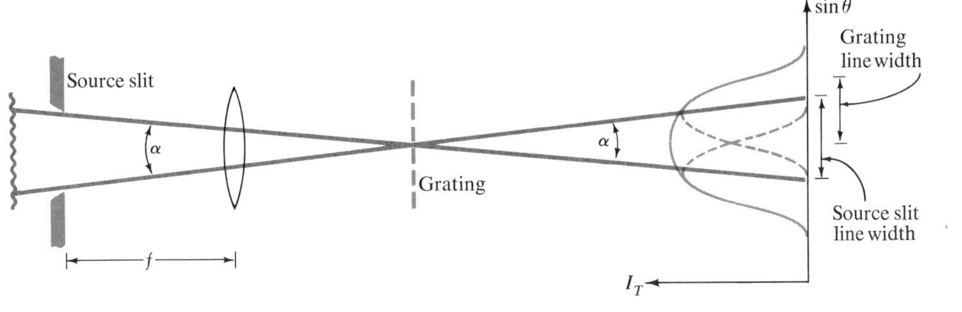

Figure 9.9: Line broadening due to a finite source slit.

angular width of this slit is added to the line width. The $m = 0$ maximum, for instance, is centered on the line from source to grating center. There is a spread of sources in a real slit; so the maxima are spread also.

Spectral lines are not infinitely narrow, even aside from instrumental effects. One source of line width is the *doppler* effect: all the atoms in the gas are in motion, since the gas is hot, and some move toward the observer while others move away. These moving sources emit waves that are doppler-shifted, some toward the blue and some toward the red, as indicated in Figure 9.10. Naturally, then, the hotter the gas, the wider the line.

Figure 9.10: Doppler broadening.

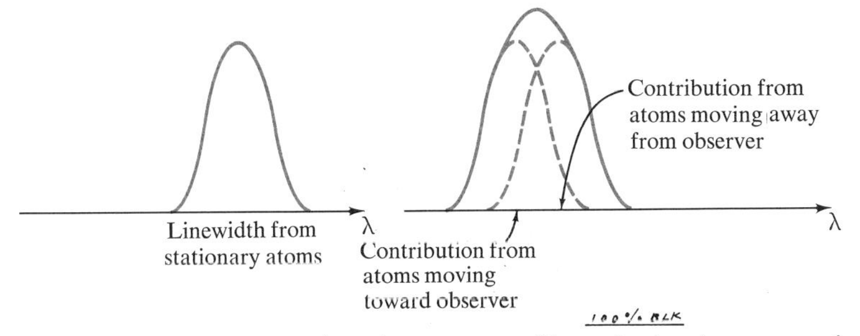

Another source of broadening is *pressure* in the gas. This is most easily thought of as due to the Uncertainty principle: $\Delta x \, \Delta p \gtrsim h$.

The way in which this works is that the higher pressures reduce the space in which the atom can wander so that the uncertainty in momentum increases, requiring larger velocities (p can never be less than Δp). These, in turn, result in larger, nonthermal doppler shifts and in broader lines.

Because spectral lines have their source in transitions between the quantum levels of atoms, we can apply the Uncertainty principle in another way: if the energy of a level is within ΔE of some value, then the photon emitted from it has an uncertainty in frequency of $h\,\Delta\nu = \Delta E$. ΔE is related to the time the atom can stay in that energy state (the "lifetime" of the state) by the Uncertainty principle: $\Delta E\,\Delta t \gtrsim h$. This lifetime is the quantity that we have already identified as the coherence time, T_c. In the next chapter we will discuss how to measure it; we will then see that the result from the Uncertainty principle could have been inferred from observation. The world standard of length is the wavelength of an orange line in the spectrum of 86-krypton, which involves a transition from a very long-lived level and is consequently extremely narrow (if all the other broadenings are minimized). *Lifetime* broadening is said to be "intrinsic", since it cannot be modified.

EXERCISES

1. Find the dark fringes from a three-slit system in which the middle slit is a source differing in phase from the other two by π.

2. Sketch the intensity pattern from a three-slit system in which the incident plane-wave fronts make an angle α with the line of the slits.

3. In Figure 9.5, if the third set of slits had been Nd apart, what would have been the relationship of the first and third sketches?

4. Show that the first secondary maximum in Figure 9.4 is about 1/22nd as high as the principal maximum (assume it occurs halfway between zeros).

5. How many slits are needed to resolve the doublet (two spectral lines) which is the major feature of the sodium spectrum? In which order?

6. Estimate the pressure necessary to make the lines of the mercury spectrum 10 Å wide (about as they are in street lamps).

PROBLEMS

9.1 Sketch the $I_T(\theta)$ curve for four slits, using the methods outlined on page 119.

9.2 Sketch the $I_T(\theta)$ curve for five slits, using the methods outlined on page 119.

9.3 Parallel light from a single source is incident on three very narrow, equally spaced slits, A, B, and C. The light passes through the slits and falls upon a screen. The intensity pattern has a central maximum at O and the intensity there is I_O. At some other point on the screen, say P, the intensity is I_P.

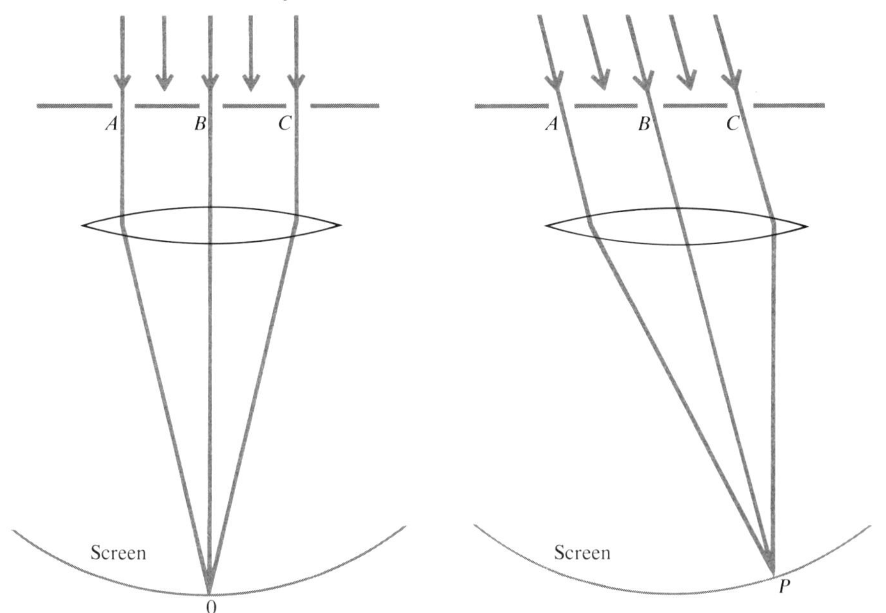

(a) If $I_P = 0$, what must be the phase difference between the light wave that arrives at P from slit A and the light wave that arrives at P from slit B?

(b) If the light wave that arrives at P from slit A and the light wave that arrives at P from slit B have a phase difference of π radians (180°), then what is the value of I_P/I_O?

(c) If P is the position of the first principal maximum, what is the value of I_P/I_O?

(d) If we denote the average value of the light intensity over the entire screen by I_{av}, then how does I_{av} compare with I_P?

9.4 In Problem 9.3, slit B is made twice as wide as A and C. Answer the same questions as in Problem 9.3.

9.5 A cube of side d has isotropic scatterers at each corner and is illuminated by a plane wave of wavelength λ, which travels parallel to one edge. If constructive interference is observed at $\theta = 0$, find the positions of eight equally bright spots. Ignore the unscattered light.

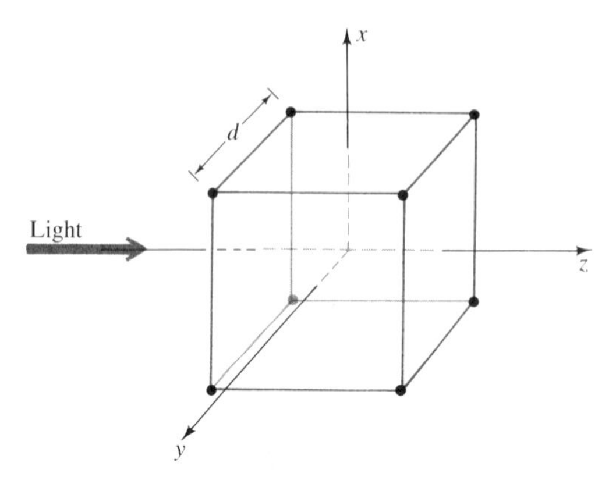

9.6 Show that the general N-slit intensity equation yields the curve of Figure 9.2 when $N = 3$.

9.7 A grating 1 cm wide has 4000 lines/cm. Half of it is illuminated by a line source S at the focus of lens L_1. Spectra are recorded on a film F at the focus of lens L_2. The line source emits light of two wavelengths $\lambda_1 = 6400.0$ Å and $\lambda_2 = 6400.5$ Å, and the focal length of L_2 is 1 m.

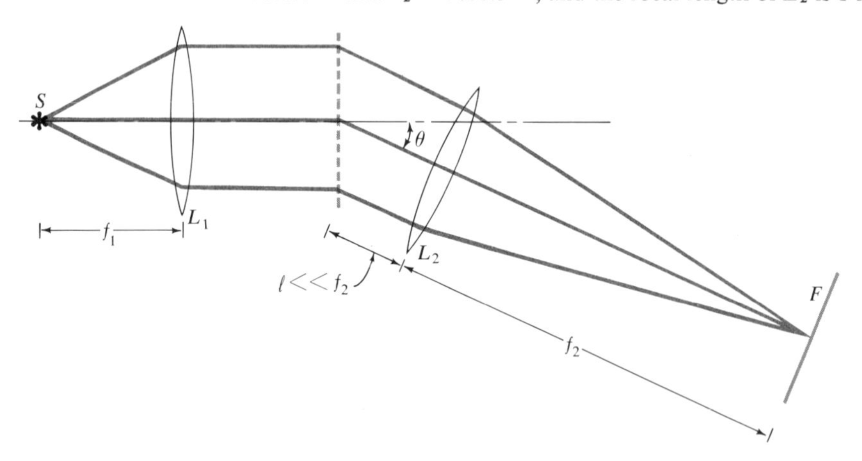

(a) At what angle θ is the third-order spectrum of light of wavelength λ formed?
(b) What is the distance between the third-order spectra of λ_1 and λ_2 on the plate P?
(c) Are these two wavelengths resolved in third order? Give reasons.

9.8 A grating of 10,000 lines, 2 cm wide, is illuminated by light of wavelength λ from a slit of width 0.5 mm, 1 m away. If the source contains wavelengths $\lambda_1 = 6400$ Å and $\lambda_2 = 6401$ Å, in what order of the spectrum will they be resolved?

9.9 A radio station has three antennas, separated by $2\lambda/3$ and driven in phase. If any one were used alone, it would radiate energy equally in all (horizontal) directions. How narrow (in angle) is the "beam" broadcast by this array? How much energy is emitted through the three antennas along the line xx?

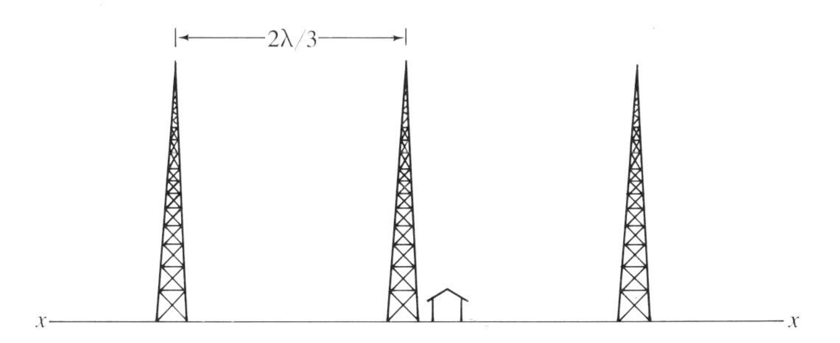

9.10 A grating has 11 narrow slits, each separated by a distance d. The curve below shows the intensity pattern as a function of $\sin\theta$ when the eleventh slit is covered up.

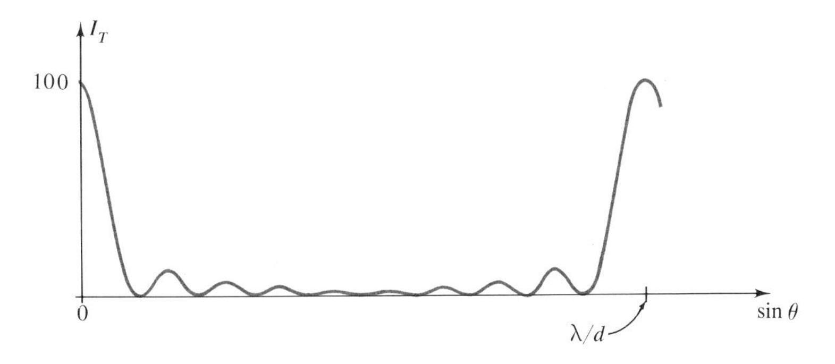

(a) All except the first two slits are covered up. Sketch the intensity curve to the same horizontal scale. Label your vertical axis properly.

(b) All except the first and last (eleventh) slits are covered up. Sketch the intensity curve to the same horizontal scale. Label your vertical axis properly.

(c) All except the first four slits are covered up. Sketch the intensity curve to the same horizontal scale. Label your vertical axis properly.

10

Multiple images: Interference of light from an extended source

We now consider an entirely different way of generating the two coherent sources which we need for interference. In this method, called *amplitude separation*, we form two or more images of a source and let the waves from these images interfere.

10.1 *Amplitude separation*

Consider first an isotropic point source in front of a mirror. The source has an image in the mirror, separated from it by a distance $2d$ and necessarily coherent to it, since both have the same phase. As we look at these sources, we see that one is farther away than the other, so we expect interference. (10.1–10.2)

We will find a bright fringe for all angles for which $\Delta L = m\lambda$. From Figure 10.1 we see that this is $m\lambda = 2d \cos \theta$. The angle θ is

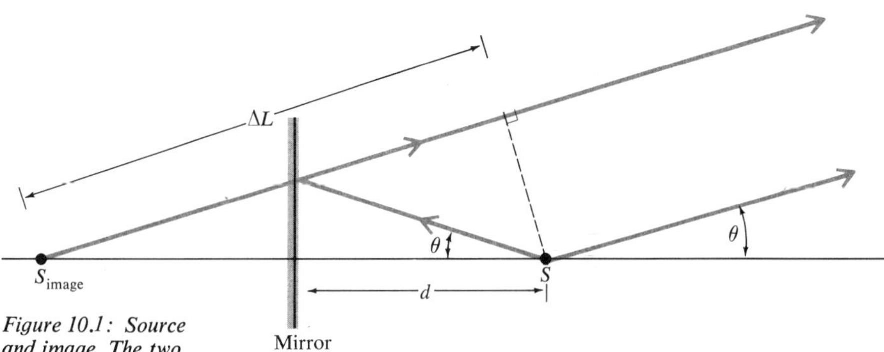

Figure 10.1: Source and image. The two are coherent.

between the observer and the line joining the sources. The "fringe", then, is a circle, since we can rotate the figure about the line joining

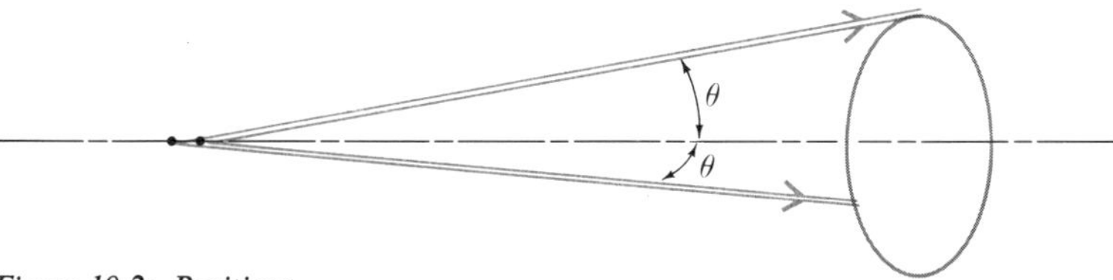

Figure 10.2: Positions where an observer sees the source-image pair as bright.

the sources and still keep θ the same. Such rings are known as "fringes of equal inclination". (10.3)

Now let us move the source (and its accompanying image) about, keeping the observer fixed at P. When $2d \cos \theta = m\lambda$, there is a bright fringe at P (the sources appear bright to the observer). This condition is satisfied whenever the source lies on the ring specified by the angle θ. From Figure 10.3 we see that, in fact, other sources on this circle also appear bright to the observer at P, although there is no interference between the total wave from the pair AA' and that from

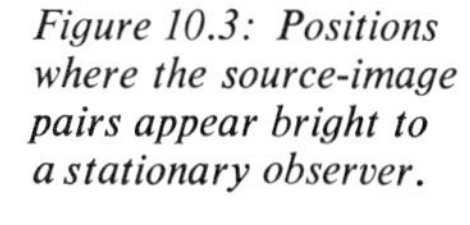

Figure 10.3: Positions where the source-image pairs appear bright to a stationary observer.

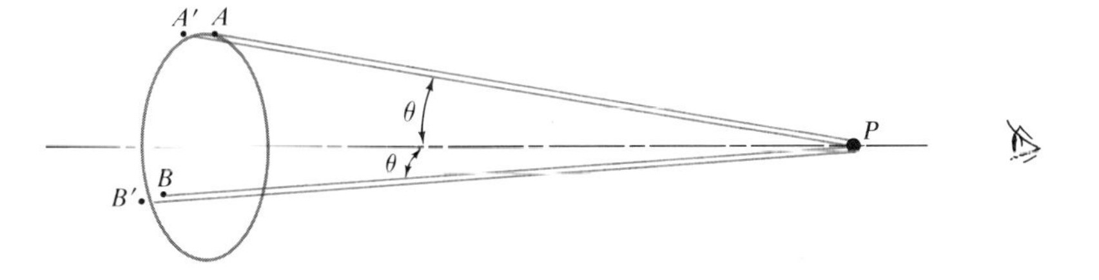

BB'. Since A and B are incoherent with respect to each other, the intensities add:

$$I(P) = I_A + I_B = (E_A + E_{A'})^2 + (E_B + E_{B'})^2.$$

This allows us to use an *extended source* for this kind of interference. That is, the source may be of fairly large extent, even when compared to its distance from the observer. What the observer sees is a pattern of bright rings. Although the source is incoherent point

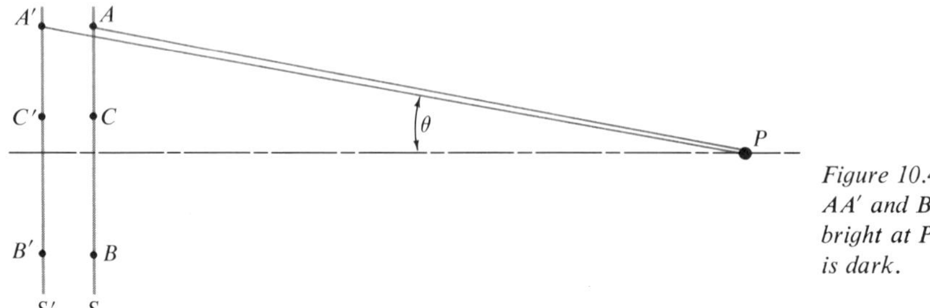

Figure 10.4: Pairs AA′ and BB′ appear bright at P. Pair CC′ is dark.

to point, each point is coherent with its image point. Referring to Figure 10.4, we see that the rings appear bright because the intensity from AA' is more than double the intensity from A. Energy is conserved: no energy goes from AA' in the direction $2d \cos \theta' = (m + \frac{1}{2})\lambda$, as it would from A or A' alone. Typically, the observer does not see the pair CC', for which P is on a dark fringe.

The advantage of this system is that we avoid the loss of intensity attendant on the use of slits, pinholes, etc. We use a large enough area of the source to make up for the intrinsic low intensity of monochromatic sources.

10.2 *Michelson interferometer*

Now it turns out to be easier to use two images of an extended source than to use the source and one image. Thus the most common two-image "interferometer" is that due to Michelson, and the typical many-image one is the Fabry-Perot. Both use "beam splitters" (generally half-silvered mirrors) to achieve the amplitude separation.

The Michelson instrument is used for very accurate measurements of distance and to measure coherence lengths ($L_C = cT_C$), while the Fabry–Perot interferometer is used in spectroscopy because its fringes can be much narrower than those of gratings. (10.4–10.6)

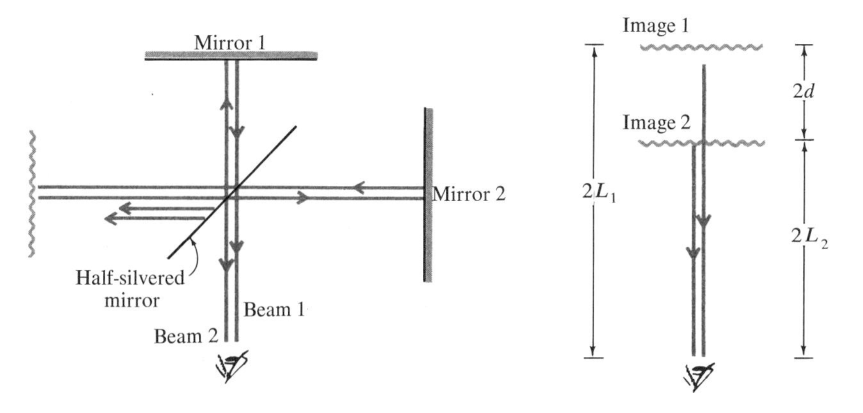

Figure 10.5: Michelson interferometer. On the right are shown the two virtual images of the source. These are coherent and appear separated by 2d.

To evaluate coherence, the Michelson interferometer, with one of the mirrors movable, is set to larger and larger values of d. When $2d \geq cT_C$, the wave from one image has changed phase before that from the other image arrives. Thus we need only look for the value of d at which the interference pattern becomes indistinct, and we will have measured the coherence length and therefore can calculate the coherence time. As d increases, the two coherent waves overlap during less and less of the observation time, so that the transition to

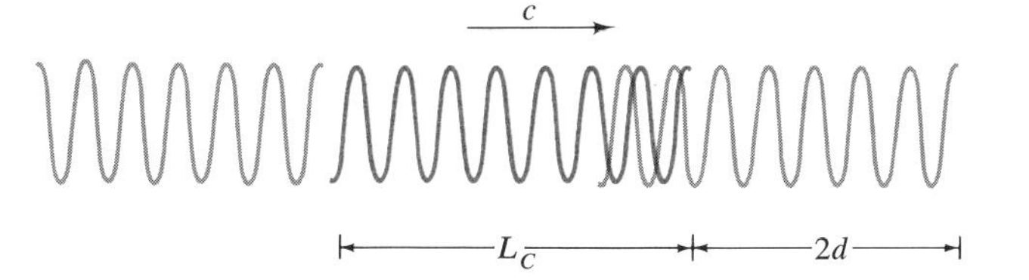

Figure 10.6: The two amplitude-separated parts of the wave train must overlap for interference to be detectable.

incoherence is a gradual one. Figure 10.6 shows a situation in which an interference pattern is visible for about 25 percent of the time. It looks visible but weak to a detector of long T_d.

The interference pattern can become indistinct for what appears at first to be another reason: the light may not be monochromatic. The situation in such a case is identical to that with the fringes from two slits: as we get far from the central fringe, where $d = 0$, the fringes due to the various colors begin to overlap and the pattern washes out. Of course, the narrower the line, the more fringes are visible (that is, the larger m may be, and hence the larger d may be). If we simply assume that the loss of the interference pattern at large d is due to an irreducible, finite, line width, then we have exhibited the inverse relationship between lifetime and frequency spread (T_C and $\Delta\lambda$), which we also derived from the Uncertainty principle. Appendix H discusses this more generally and points out its mathematical implications.

10.3 *Fabry–Perot interferometer*

The Fabry–Perot interferometer is a multiple-source device. The two mirrors face each other as illustrated in Figure 10.7, so that there is an infinite series of images of the source, each separated by $2d$. Just as the N-slit grating gave us lines N times narrower than those of

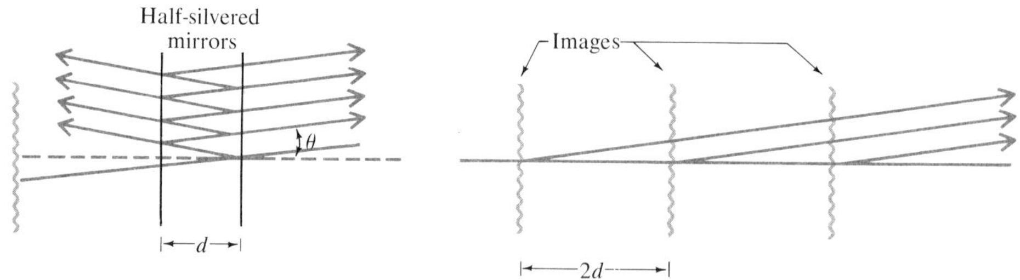

Figure 10.7: Fabry-Perot interferometer. On the right are shown some of the infinite number of (coherent) virtual images.

the two-slit device, so the Fabry–Perot gives us narrower fringes than those given by the Michelson instrument. In the case of the Fabry-Perot, each successive image is slightly less intense, so the phasors gradually shrink. The comparison is sketched in Figure 10.8. Analysis of this instrument is difficult, but it seems reasonable that the more

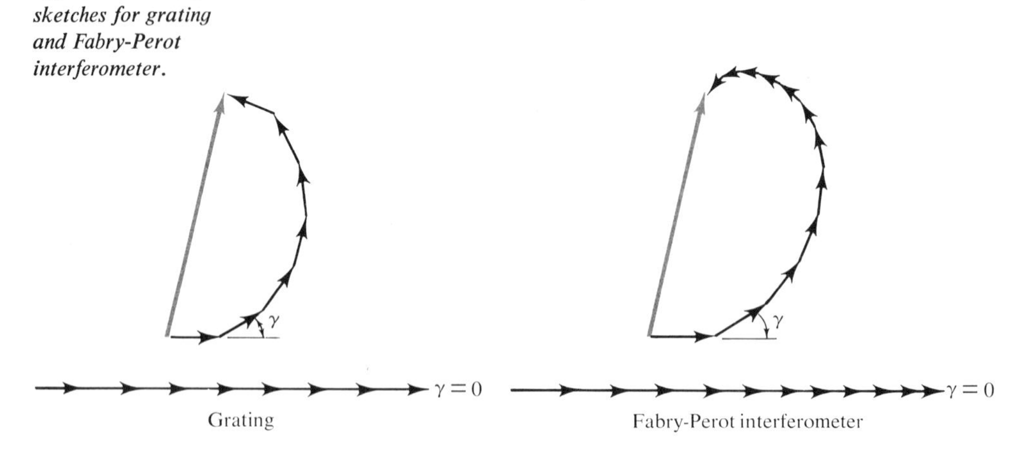

Figure 10.8: Phasor sketches for grating and Fabry-Perot interferometer.

the mirrors are silvered, the more images there will be and consequently the narrower will be the fringes. Fabry–Perot interferometers are used for very high resolution "spectroscopy": doubled fringes mean doublet spectral lines, which a grating might not resolve.

For instance, a close doublet may contain wavelength λ and $\lambda + \delta\lambda$, where $\delta\lambda$ is so small that no grating resolves the pair. But for a Fabry–Perot interferometer set at large d, the separation of the fringes may well be larger than their width. Typically, doublets of separation 0.01 Å are resolved this way. Such doublets often arise from the existence of several isotopes of a given element. Since the nuclei have different masses, the spectral lines are shifted slightly. Similarly, the intrinsic magnetic moments (spins) of electrons and some nuclei lead to small "hyperfine" splittings of spectral lines, resolvable with Fabry–Perot interferometers. Much of our detailed knowledge of atoms comes from high-resolution "spectroscopy", which is mostly done with interferometers. (10.7)

10.4 *Wedge*

An example of the kind of interference we are discussing here is provided by the "wedge" of Figure 10.9, two partially reflecting surfaces inclined at a small angle to each other. If the angle is small, the normal to one is nearly the normal to the other. We can keep the

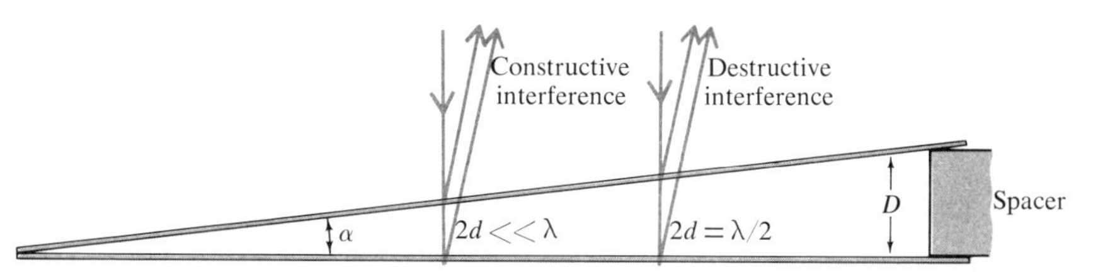

Figure 10.9: Reflecting wedge.

problem simple by observing nearly along these "normals", so that $\cos\theta \simeq 1$. But since the distance d varies along the wedge, the fringe at $\theta = 0$ changes from dark to bright every time $2d$ increases by another half-wavelength. Since $2d = 2D\tan\alpha$ and it is easy to measure D, this provides an extremely sensitive way of measuring α. Alternatively, it is a sensitive way of measuring the thickness of the spacer which is used to form the wedge. (10.8; 10.9)

A common kind of wedge is that formed by the departures from flatness of a surface. As an example, suppose two mirrors are in contact along their edges, but bow apart near the center, as shown in Figure 10.10. As we look farther from the edges, the separation increases until $2d = \lambda/2$. At this point, one reflected wave interferes destructively with the other, and we see a dark fringe. Farther toward

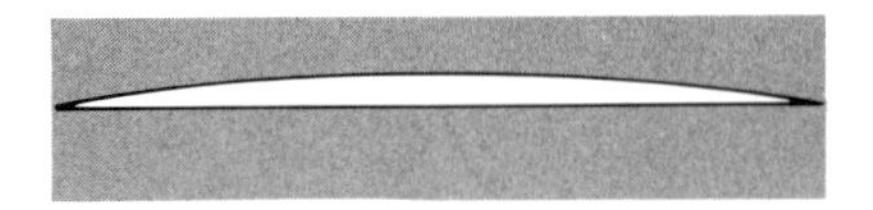

Figure 10.10: Two nonflat surfaces form a kind of wedge.

the center, d increases more, and when $2d = \lambda$, a bright fringe again occurs. Each new bright fringe which appears means a further increase of λ in $2d$. So if we count six fringes from edge to center, the total departure from flatness is $2d_{\text{max}} = 6\lambda$ or $d_{\text{max}} = 3\lambda$. This is so convenient that optical mirrors are often said to be *flat* to a given number of fringes.

An interesting kind of wedge is formed when a spherical surface like that of a lens is in contact with a plane surface. The circular fringes produced by this arrangement are known as "Newton's rings".

10.5 *Transmitted light*

In the example of Section 10.4, we viewed the two mirrors by reflected light. In fact this can be difficult if the amplitudes of the two waves differ greatly. Then the destructive interference reduces E_T, but never to zero. For instance, if $E_{\text{trans}} = \frac{1}{2}E_{\text{reflect}}$ at a surface, the beams have the amplitudes shown in Figure 10.11.

Thus, considering only the first two beams (A and B in Figure 10.11), the ratio is 9 : 1, so that E_T varies from $(10/27)E_0$ to $(16/27)E_0$. Further reflections improve this somewhat, but the contrast is always

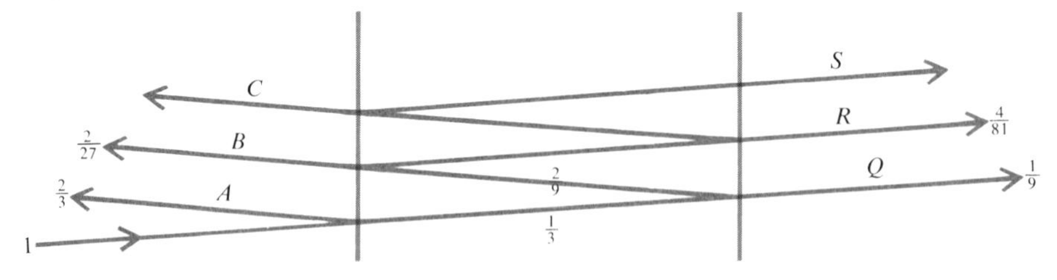

Figure 10.11: Amplitudes of transmitted and reflected waves will determine the better way to observe the interference.

poor. In comparison, the first two transmitted beams (Q and R) have the ratio amplitude 9 : 4, with consequent improvement in contrast.

This means that we will want to view silvered wedges by transmission. The student can easily show that an unsilvered wedge ($E_{\text{trans}} = 5E_{\text{reflect}}$) should be viewed by reflected light.

10.6 *Phase change on reflection*

So far we have neglected one aspect of the real devices that give rise to our multiple images. We have supposed that the wave is changed in amplitude, but not in phase, at every reflection. Now we must modify this assumption to explain the following: nothing in our devices absorbs energy, but at certain angles no energy appears. When discussing wave-front division (slits), we explained this as being due to the waves propagating somewhat sideways, according to Huygens' principle. But here we have no nearby boundaries, so the explanation is something else. We find experimentally that when we see a dark fringe in the transmitted pattern, we see a bright one in the reflected light. This is because one of the reflections involves a phase change of π.

It turns out that an *external* reflection (light from a medium of low index of refraction reflected from the interface with a medium of higher index) changes the wave's phase by π. An internal reflection leaves the phase unchanged. This is the same situation we had with waves on a string, that is, a phase change of π at a fixed end and no change at a free end.

It seldom matters whether we remember which kind of reflection gives the phase change, as long as we remember there is a difference. For instance, the reflected beam in our parallel mirrors involves two

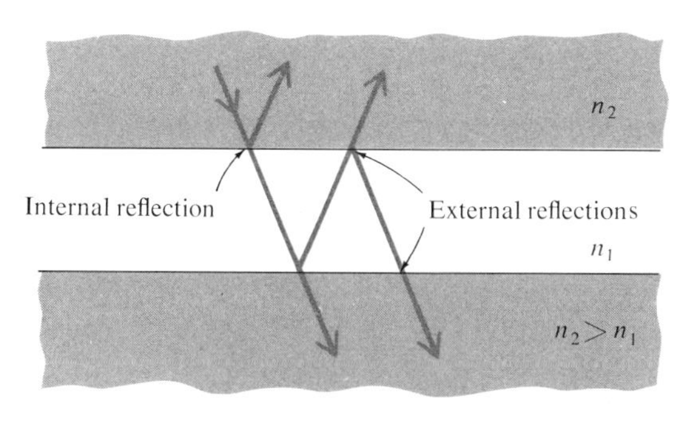

Figure 10.12: The reflected and transmitted waves differ in the number of phase-changing reflections they undergo.

reflections, one of cach kind. Both transmitted beam reflections are of the same kind. So the reflected beam has an extra *net* phase change of π, as shown in Figure 10.12. Thus the condition for a bright fringe for the transmitted beam is $\Delta L = m\lambda$, since the extra half-wavelength is necessary to make up for the phase change. The same is true of a single slab of glass, like that of Figure 10.13. Notice that all *further*

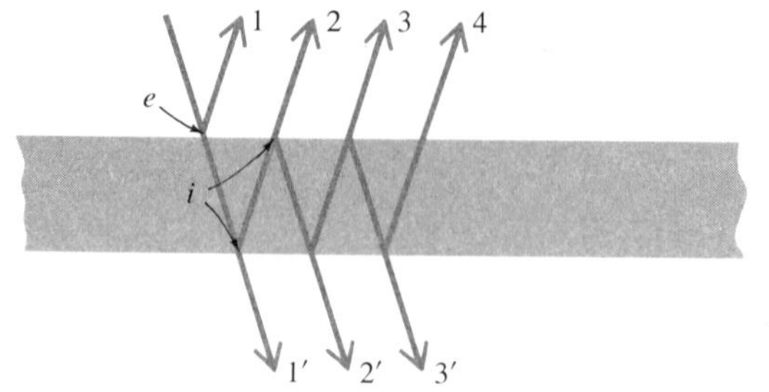

Figure 10.13: Successive waves contribute to the interference in a way to reinforce the situation found with the first two.

reflections have an even number of phase changes so that if ray 1 interferes destructively with 2, it does so also with 3, 4, 5, etc. It is always true that

$$E_1 \geq \sum_{i=2}^{\infty} E_i .$$

That is, E_1 is never less than the sum of all other amplitudes (at that angle and value of d).

10.7 *Wavelength dependence*

In most applications of interference from films we are concerned with only one wavelength. To see the effect of varying λ, let us consider two images of a source separated by $2d$, when d is very small. Then, as long as θ is reasonably small, we can assume that we are viewing the central spot ($\theta \cong 0$). (For instance, if $d = \lambda/2$, we have a bright fringe ($m = 1$) all the way out to $\cos\theta = \frac{1}{2}$. Thus the film appears bright for all viewing angles within 60 degrees of the normal.) Now suppose that the wavelength in question was 6000 Å. If we were to illuminate the same film with bluer light, say at 4000 Å, the appearance would be different. At 4000 Å, the thickness is $\frac{3}{4}$ wavelength so that the central fringe is dark (and extends out to $\cos\theta = \frac{2}{3}$, $\theta = 48°$). If we illuminate with both wavelengths, only the red is seen. Such a situation is shown in Figure 10.14.

If we apply our example to white light, we find that we have actually built a filter that lets the red light through preferentially. In the problems it will be seen that a thicker film has a narrower "pass-band", but is more directional. More usefully, perhaps, we

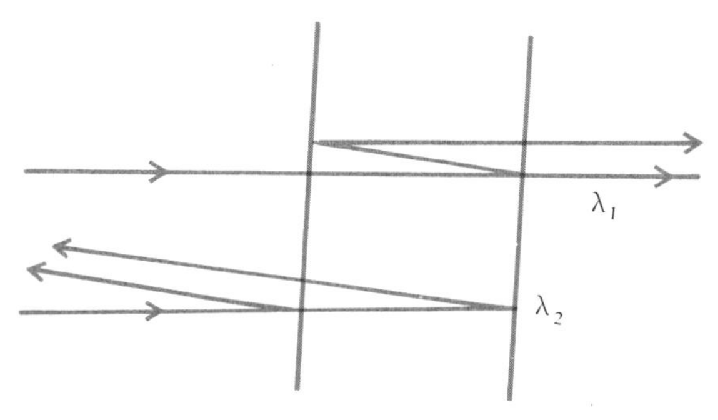

Figure 10.14: The condition for a bright fringe on transmission for λ_1 may correspond to a dark fringe for λ_2.

may silver the film and get a very narrow spread of wavelengths for which there is a bright fringe, retaining the directional benefits of the thinner device. Such an arrangement is called an *interference filter*. (10.10–10.12)

The colors seen in soap bubbles and oil slicks are interference colors. Notice that if the medium separating the two images is not a vacuum, the condition for constructive interference is that the path difference of the two beams must be equal to an integral number of wavelengths in the medium. Therefore the more general condition for a bright fringe is $2nd \cos \theta = m\lambda$.

EXERCISES

1. A line source is reflected from two parallel mirrors. Regard this as a two-slit problem and find $I(\theta)$.

2. If the separation of the apparent sources in a Michelson interferometer is $d = 0$, a uniform bright field is seen. Then one mirror is rotated through the angle α. Describe the new pattern (quantitatively).

3. The plates in a Fabry–Perot interferometer are silvered to give $E_{\text{reflect}} = \frac{1}{2} E_{\text{inc}}$. Use a phasor sketch to estimate the fringe width.

4. A flat piece of glass 200 in. across is covered with a thin layer of water. Find the number of fringes from edge to center due to the curvature of the earth, which the water surface follows.

5. Find the colors of the light transmitted by a soap-water film 2 μm thick.

6. Find the angular position of the first color reversal for the film of Exercise 5.

PROBLEMS

10.1 A single dust particle acts like an isotropic scatterer. Such a particle is on the front (unsilvered) surface of a mirror so that one may observe the light scattered directly from the particle and that scattered to the mirror and back. Find the observed interference pattern.

10.2 A point source is reflected by the two surfaces of a flat piece of glass. What pattern is observed?

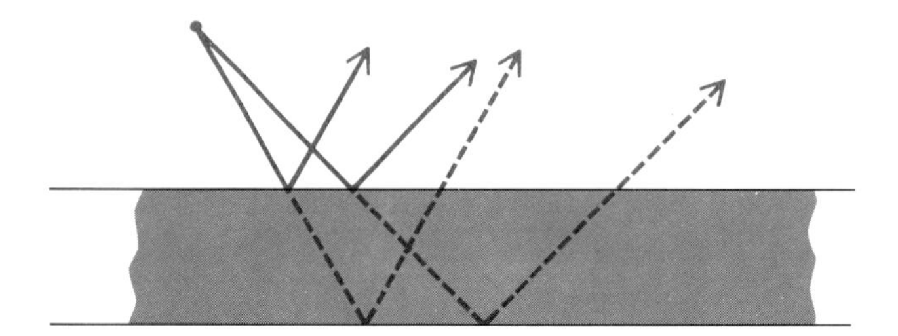

10.3 Find the apparent position of the fringes of equal inclination: of the fringes crossing the wedge. By apparent position we mean the place where the eye is focused to see them.

10.4 A thin film with $n = 1.40$ for light of wavelength 5890 Å is placed in one arm of a Michelson interferometer. If this causes a shift of seven fringes, what is the film thickness?

10.5 A Michelson interferometer is used to view a source whose light falls in the wavelength range λ_0 to $\lambda_0 + \delta\lambda$, where $\lambda_0 \gg \delta\lambda$. Estimate the range over which the movable mirror can be moved without loss of visibility of the fringes. What is the coherence time of this light?

10.6 The apparatus shown is a Mach–Zender interferometer. Both beam splitters transmit 90 percent of the light intensity incident on them and reflect the other 10 percent. Mirror 2 is not parallel to the others, by the angle δ.

(a) What pattern does observer A see?

(b) What pattern does observer B see?

(c) If a half-wave plate with its fast axis vertical is inserted in the beam at C, and one with its fast axis horizontal is inserted in the beam at C', what does the observer A see?

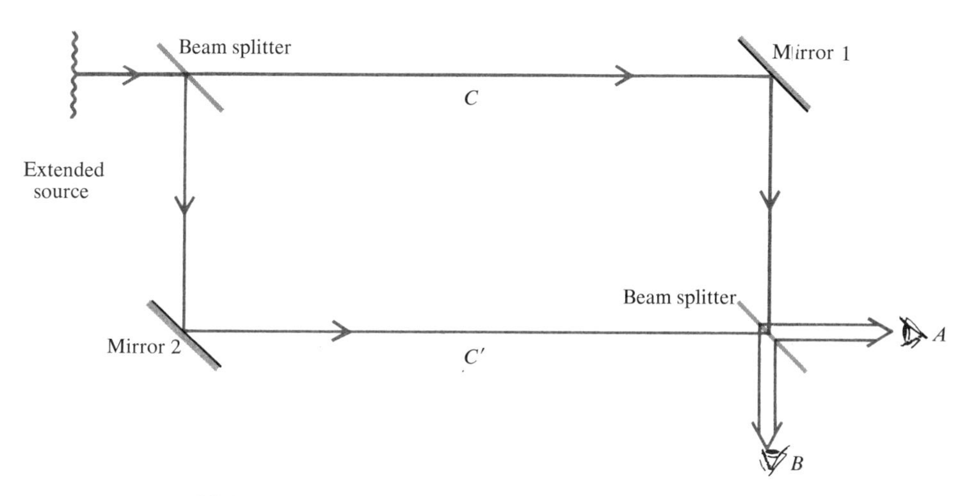

10.7 A Fabry–Perot interferometer of spacing 1 mm just resolves a doublet spectral line. Find $\Delta\lambda/\lambda$.

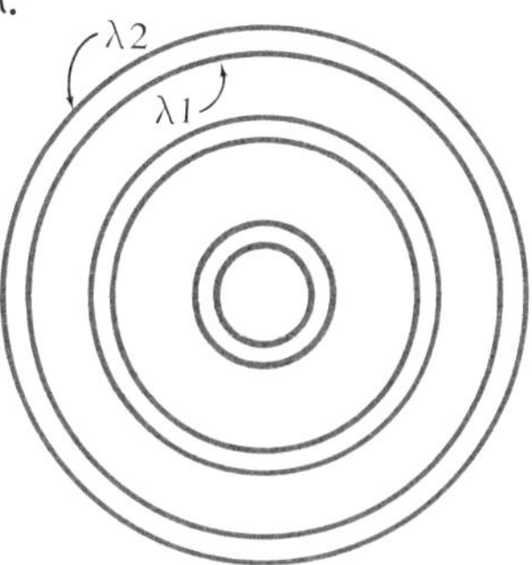

10.8 The two flat plates shown are in contact at one edge and are separated at the other by a spacer 5 μm thick. The upper plate has an index $m = 1.5$; the lower, $n = 2.0$. Fringes are observed in reflected light of wavelength 0.5 μm.

(a) How many bright fringes are observed? What sort of fringe is at the contact point? Near it?

(b) If the whole works is immersed in oil of index $n = 1.8$, answer the same questions as in part (a).

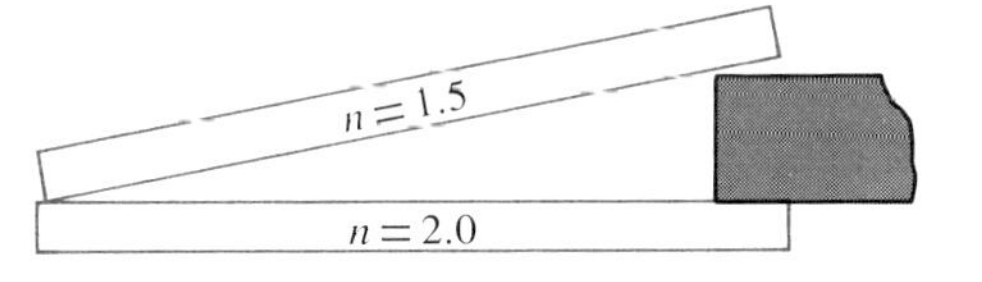

10.9 The wedge shown in (a) of the figure is made of two flat glass plates, of index 2 and 4/3. The material between them has index n_0. Viewed in reflected light of frequency ν, the fringe pattern looks like that shown in (b).

(a) Give limiting values for n_0. That is, ($n_0 < ?$) and/or ($n_0 > ??$).

(b) *Estimate* the thickness of the spacer.

(c) Sketch the intensity pattern if n_0 is doubled.

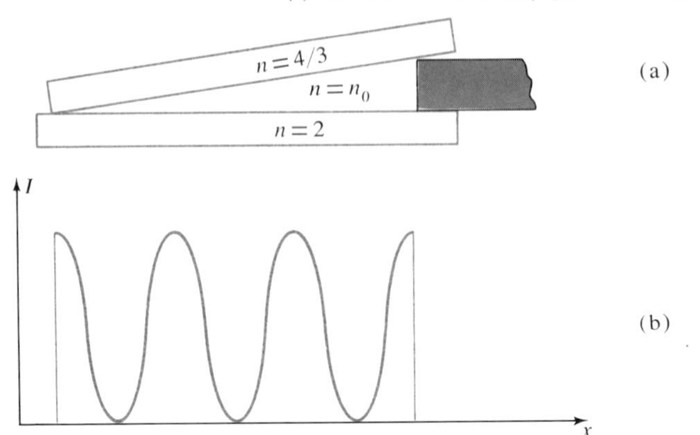

10.10 White light is reflected from a thin film of thickness d and index n, and then enters a grating spectrometer. Find the spacing in wavelength of the dark bands ("channels") crossing the observed spectrum.

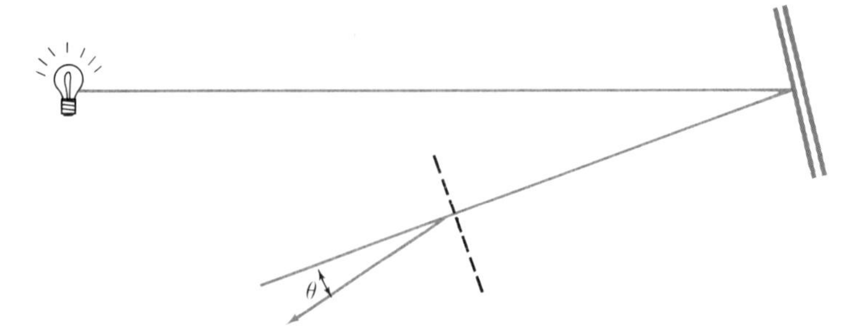

10.11 A plane wave of monochromatic light falls normally on a uniformly thin film of oil which covers a glass plate. The wavelength of the source can be varied continuously. Complete destructive interference of the reflected light is observed for wavelengths of 5000 and 7000 Å and for no wavelengths between. If the index of refraction of the oil is 1.30 and that of the glass is 1.50, find the thickness of the oil film.

10.12 In the situation of Problem 10.11, find the dependence on angle of the light transmitted at 6000 Å.

11
Diffraction

We now come to the study of diffraction, a phenomenon akin to interference, but not depending on multiple sources. Historically, diffraction was observed before interference, and it is still more easily seen with minimal apparatus. In optics, the phenomenon is this: The edges of shadows (for instance, those around the image of a slit) are not sharp, but have "fringes" of light and dark along them. The origin of the effect is similar to the origin of fringes in interferences, but the calculation can become much more complex. We will start by treating the case of a single diffracting slit in some detail, since it is a simple one to discuss.

11.1 *Single slit*

Figure 11.1 shows a slit illuminated by a plane wave. According to Huygens' principle we can think of every point on the wave front as

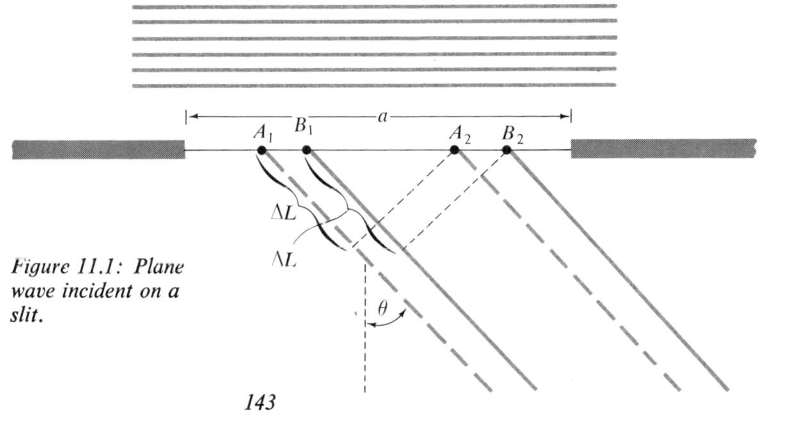

Figure 11.1: Plane wave incident on a slit.

a (point) source of a new wavelet. To find the first dark fringe, we divide the slit into two parts. Every point in the right half has a partner point in the left, a distance $a/2$ away. When one of the points of a pair is $\lambda/2$ farther from the observer than the other, then the net **E** vector reaching the observer from the pair is equal to zero. If this situation holds for one pair, then it must for all, in the Fraunhofer limit. Thus the light intensity is zero and we observe a dark fringe. This situation holds when $\Delta L_{\text{pair}} = \frac{1}{2}\lambda = \frac{1}{2}a \sin\theta$. That is, $a \sin\theta = \lambda$ specifies a dark fringe.

There are other dark fringes derivable from this set of paired points. They occur at $\Delta L_{\text{pair}} = (m + \frac{1}{2})\lambda = \frac{1}{2}a \sin\theta$; that is, at $a \sin\theta = (2m + 1)\lambda$. But these are not all the possible fringes. For instance, if we divide the slit into four equal parts, the light from points A and C might be in phase, but canceled by that from points B and D. This situation leads to

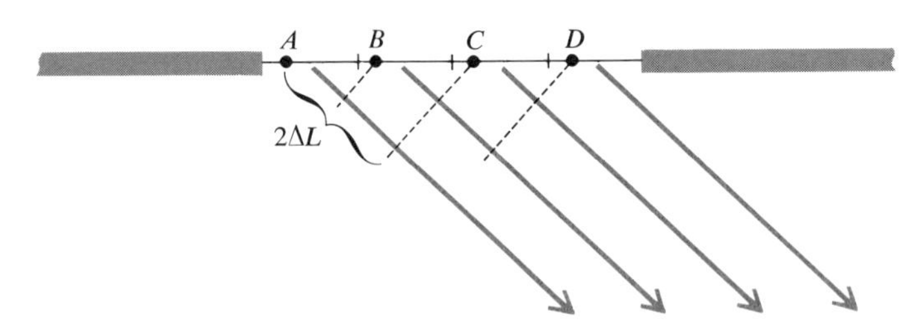

Figure 11.2: Division of the slit into four equivalent parts.

$$\Delta L_{(AB)} = \frac{\lambda}{2} = \frac{a}{4} \sin\theta,$$

which means that a dark fringe occurs when $a \sin\theta = 2\lambda$. We can continue to play this game until we find all the dark fringes. They are represented by:

$a \sin\theta = m\lambda$ dark fringe.

The student should convince himself that no others occur. For instance, it is easily shown that division of the slit into three parts is redundant, yielding the fringes already found.

The bright fringes do not come halfway between the dark ones. For instance, the point halfway between the first and second dark fringes is specified by $a \sin\theta = 3\lambda/2$ or $(a/3) \sin\theta = \lambda/2$, which represents a division of the slit into three parts, with two of the waves canceling and one being left over. The bright fringes can be located, but the procedure is difficult and seldom necessary. The more useful information is that specifying the location of the dark fringes.

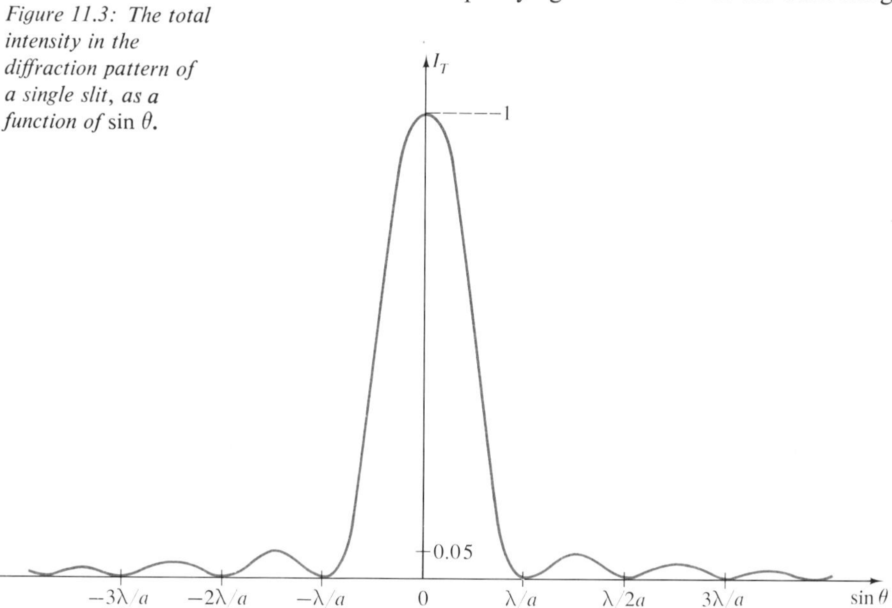

Figure 11.3: The total intensity in the diffraction pattern of a single slit, as a function of $\sin\theta$.

As we see from Figure 11.3, successive maxima fall off quite rapidly in intensity, so that we often perceive mainly the central maximum. The width of this bright area is found in the same way as the widths of spectral lines formed by a grating: the range of $\sin\theta$ between the adjacent dark fringes is $2\lambda/a$. This is the "width" of the central bright line in the diffraction pattern of a slit. Notice that it is small when $\lambda \ll a$ (the region of geometrical optics), and spreads out as a approaches λ in size. In fact, if $a < \lambda$, no dark fringe ever appears, and we find we have constructed a "point source", whose existence we have been assuming rather uncritically all along.

11.2 *Diffraction-limited optics*

We have occasionally assumed the existence of perfect lenses. We should now take note of the fact that even if a lens has perfect surfaces, the very fact of its finite size limits the sharpness of the image. As an example of such a "diffraction limited" lens, consider how small a point we may focus a laser to. We suppose the laser emits a perfectly parallel (that is, plane wave) beam of diameter d_1, which is brought to a focus by a lens of focal length f and diameter d_2. Whichever of the two diameters is smaller, then, defines the aperture in question. We can see this by imagining an actual aperture "stopping down" a larger beam or lens, as in Figure 11.4(b), where we have used a as the aperture size.

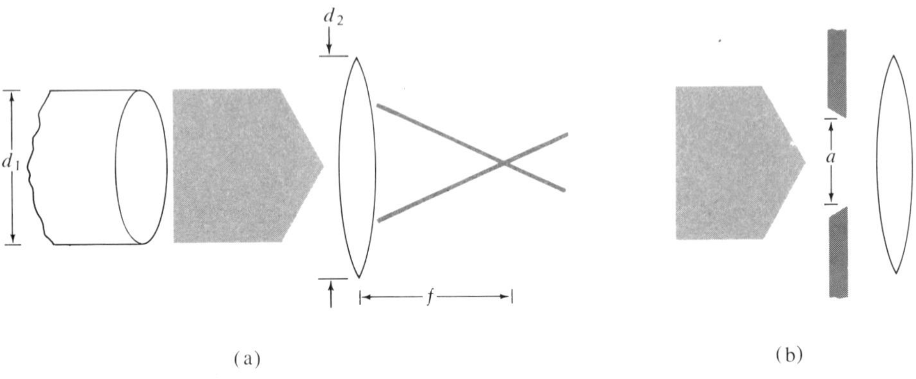

Figure 11.4: A focused laser beam, diffraction limited.

The "image" at the focal point of the lens is not a true point, but rather the diffraction pattern of the aperture. This has a central maximum of angular diameter $\theta_0 = \lambda/a$, and therefore of diameter $D = f\lambda/a$. We can use a short-focus lens, a microscope objective, say, and we can keep a as big as possible. But inevitably the spot is of finite size because even the fastest lenses have f/a of about 1, leaving $D \simeq \lambda$.

It might be of interest to estimate the electric field in such a minimal spot: a laser may emit 10 J of energy in a pulse lasting for 10^{-9} sec. This means that the energy density at the spot is about $10\ \text{J}/D^2c\ 10^{-9}\ \text{sec} \cong 10^{13}\ \text{J/m}^3 = \frac{1}{2}\epsilon_0 E^2$. So $E \simeq 10^{12}$ V/m, which accounts for some of the laser's wallop.

11.3 *Resolution*

We have discussed the resolution of spectral lines by a grating. A similar problem arises in relation to the images of small objects: because of the diffraction pattern, the image is not sharp, so that it may become difficult to tell how big an object is or to distinguish between one object and two. Again we adopt an arbitrary criterion as the definition of "resolution" of two images: if the central maximum of one pattern is no closer than the first dark fringe of the other, the images are said to be resolved — that is, there are two detectable images. That this is not an exact definition is seen from Figure 11.5(c), since to an educated eye the images are discernable as two.

Figure 11.5: Resolution of two diffraction patterns.

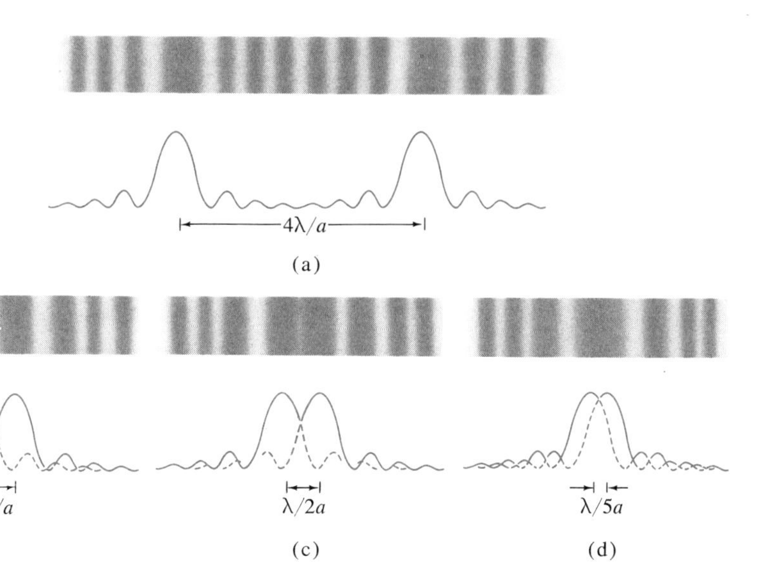

Since circular apertures abound in optical devices, it is well to mention that the pertinent value of $\sin\theta$ for round holes is

Circular aperture: First dark fringe at $\sin\theta = \dfrac{1.22\lambda}{a}$.

Straight slit: First dark fringe at $\sin\theta = \dfrac{\lambda}{a}$.

We generally formulate the discussion of resolution as if we had two slits illuminated by a single diffuse source, a situation which would pertain in looking at our double-slit apparatus with a magnifying glass, for instance. But more common is the situation in which we observe two sources (for instance, a double star) through a single aperture like the pupil of the eye or the objective lens of a telescope. Two such point sources are incoherent to each other, but each has a diffraction pattern due to the aperture. With a large aperture the patterns do not overlap, and the sources are resolved, as shown in Figure 11.6.

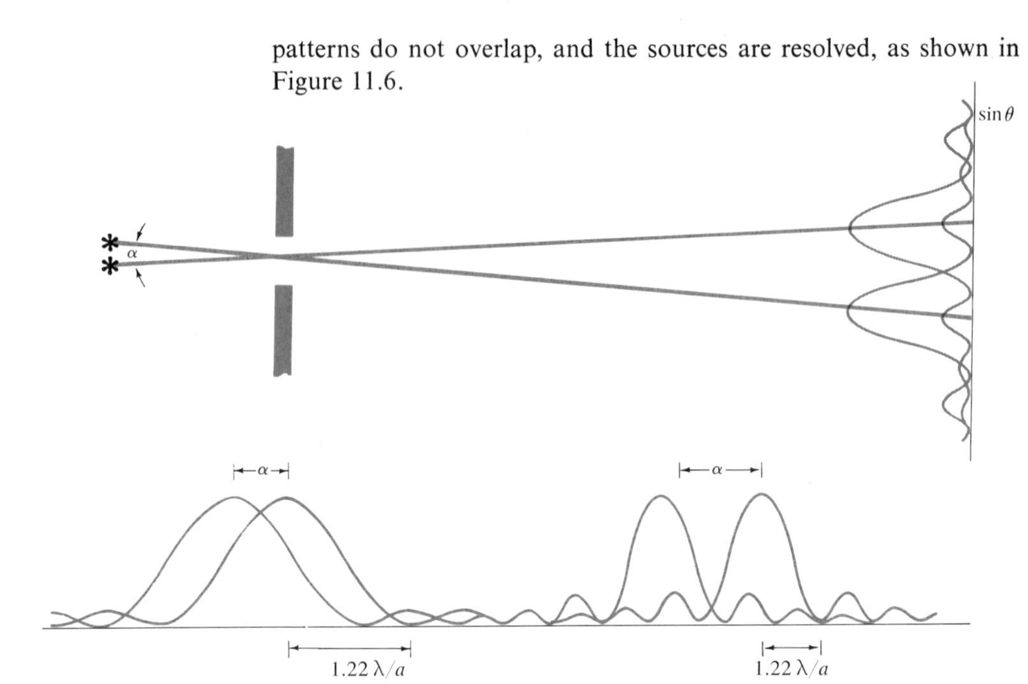

Figure 11.6: Resolution of two stars. Part (b) shows the pattern in a telescope of small aperture: part (c) the same stars seen through a large instrument.

An extended object may be thought of as many point sources, side by side, but not coherent to each other. Figure 11.7 shows the resulting diffraction pattern. A star is an excellent example of an extended object which is so far away that the angle it subtends is negligible compared with the diffraction width introduced by even the biggest telescopes.

Figure 11.7: Resolution applied to an extended object.

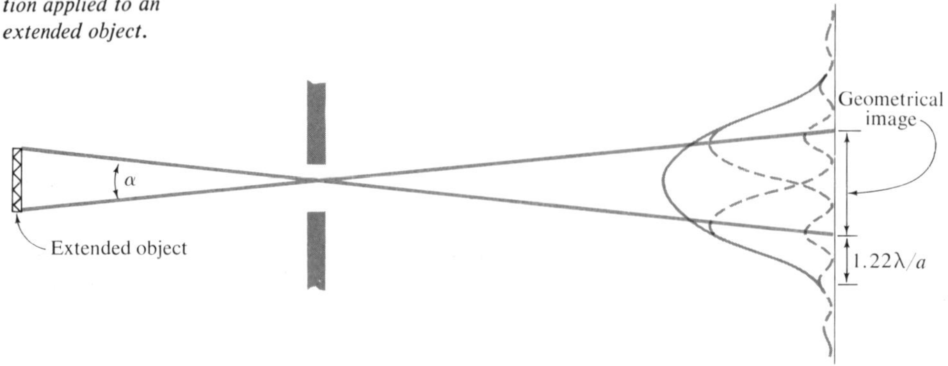

Notice that an absolute resolution limit exists, in that an aperture (or an object) of size comparable to a wavelength of light has a central maximum so wide that it overlaps all others. This means that in order to examine objects smaller than about half a micrometer, we must use a microscope which employs waves shorter than visible light. This need is met by the ultraviolet microscope or (for very small objects) by the electron microscope.

11.4 *Babinet's principle*

Small opaque objects give diffraction patterns just as apertures do. A useful relationship exists between the patterns due to an object and its complementary aperture—an aperture of the same size and shape as the object. *Babinet's principle* states that the diffraction patterns of complementary diffractors are identical in the Fraunhofer limit! A simple argument in support of this is the following: the **E** vector at some point P is $\mathbf{E}_a$ when the diffraction is through an aperture, and is $\mathbf{E}_c$ when the diffractor is the complementary object. If both the object and the screen with the aperture are placed between the light and P, no light reaches P. So $\mathbf{E}_a + \mathbf{E}_c = 0$. This means that $\mathbf{E}_a = -\mathbf{E}_c$. We square both to find the intensities $I_a = I_c$, as illustrated in Figure 11.8.

Figure 11.8: Diffraction-pattern amplitudes and intensities from a slit and its complementary object.

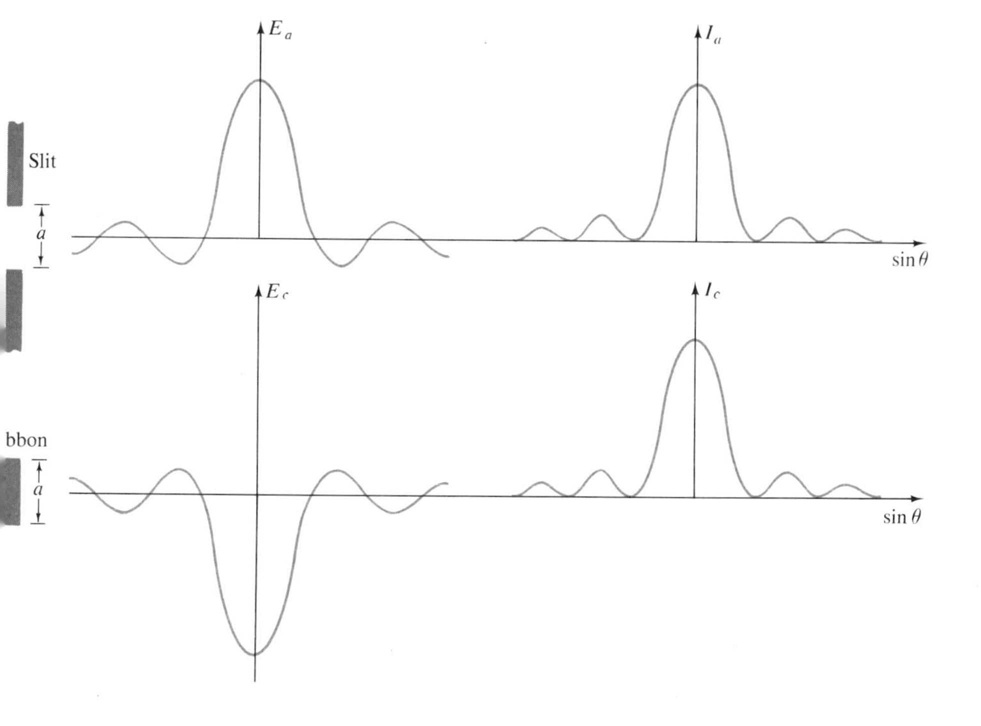

Babinet's principle is very useful in many instances, but it must be used with care. The argument just given must be modified slightly for the exact $\theta = 0$ condition, when a bright uniform background is also present in the case of the complementary object. The situation must also be modified when the Fraunhofer condition is not satisfied and in certain cases when the diffractors are metallic. But the general idea is correct, and within these restrictions is very useful.

11.5 $I(\theta)$ — *single slit*

The single slit provides us with a solvable diffraction problem. We will find an expression for the complete pattern in order to illustrate the general approach, even though a knowledge of the positions of the dark fringes is more useful in real single-slit problems. We use the same phasor construction that we used with the grating except that now each phasor has infinitesimal length and differs from its predecessor by an infinitesimal angle, which means that we know only the total length and the total angle.

If we had N phasors, each turned through an angle β_i, we would say $N\beta_i = \beta$. Here we take a limiting case of this, where $N \to \infty$, $\beta_i \to 0$, but β is unchanged. Similarly, each phasor has length E_i, and $NE_i = E_0$. Then, in the limit of $N \to \infty$, $E_i \to 0$, we keep E_0 unchanged. This corresponds to the phasors, each of infinitesimal magnitude, representing the light from the infinite number of point sources into which we can divide the slit. The amplitude of the resultant phasor is E_0 when $\sin\theta = 0$, since the light from every point on the slit arrives at the (distant) observer in phase. When $\sin\theta$

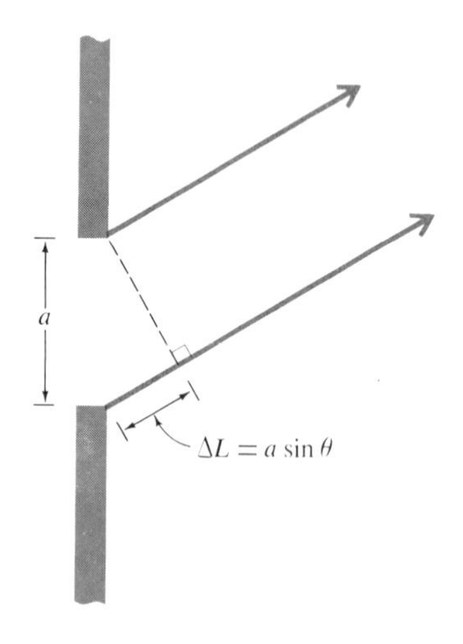

Figure 11.9: Extreme pathlength difference for a single slit.

is not zero, the observer finds light from one edge of the slit differing in phase from that at the other edge by the total phase angle β, where we see from Figure 11.9 that $\beta = (2\pi/\lambda)a \sin \theta$.

Now the resultant amplitude of all the infinitesimal phasors added together is E_T. We square this to find the resultant intensity. Notice that only the magnitude of E_T is desired, since our slow detectors do not preserve the information on its phase (relative to some arbitrary phase zero). Pictorially, this means that we can orient our first phasor in any convenient way. The recovery of the phase information will be the basis of the next chapter, where we treat the subject known as "modern optics".

The simplest situation we can calculate is that with $\sin \theta = 0$, so that $\beta = 0$. Then all the phasors are in line and $E_T = E_0$. This defines E_0 and illustrates the point made above about the phase. A second simple geometry is that of the first dark fringe, where $\sin \theta = \lambda/a \quad (\beta = 2\pi)$. In this case the resultant is zero, as illustrated in Figure 11.10.

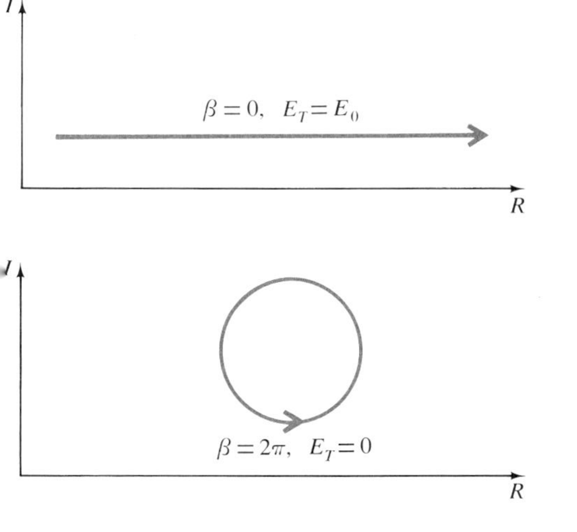

Figure 11.10: Phasors from a single slit, central maximum and first zero.

Now let us make the general calculation. In Figure 11.11 we draw the arc formed by the phasors at some general value of β, and we wish to find E_T, a chord of this arc. Draw the lines H perpendicular at each end of the arc, and a perpendicular bisector of the chord. Then, from geometry, we can write two expressions for H: $\quad \beta H = E_0$,

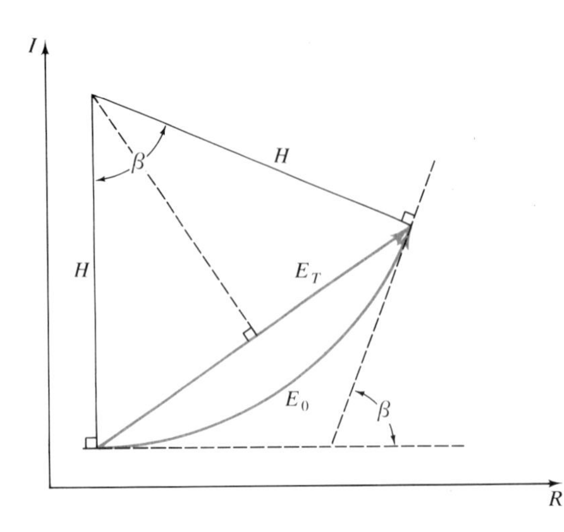

Figure 11.11: Phasor Construction used to find $I(\theta)$ for a single slit diffraction pattern.

expressing the arc length in terms of β (in radians), and $\sin(\beta/2) = (E_T/2)/H$. We then eliminate H to find

$$E_T = E_0 \frac{\sin(\beta/2)}{(\beta/2)}, \qquad \text{and therefore} \qquad I_T = I_0 \frac{\sin^2(\beta/2)}{(\beta/2)^2}.$$

It is easily verified that this does indeed have its zeros at $a \sin\theta = m\lambda$, except for the point at $\theta = 0$ where the expression has a maximum.

The extension of this procedure to more complicated geometries is straightforward, but is impossibly difficult in most cases. In the next chapter we will consider a more general way of looking at the problem, as well as some outgrowths of modern diffraction theory.

EXERCISES

1. Show that division of a slit into three parts leads to no new dark fringes.

2. Find the intensity in a single-slit diffraction pattern at $\sin\theta = 3\lambda/2a$ (halfway between first and second dark fringes).

3. What error is there in the geometrical optics assumption of straight-line light travel through a window 1 m square? That is, how well defined is the bright area relative to its total size?

4. How close may two planets be, as they near conjunction, before the human eye can no longer resolve them? (Do you believe this?)

5. Describe quantitatively the diffraction pattern of a collection of opaque spheres each 2 μm in diameter.

6. A single-slit diffraction pattern falls on a screen 1 m from the slit. Find the intensity distribution on the screen, $I_T(Y)$.

PROBLEMS

11.1 Find the intensity distribution $I(x_0, y_0)$ of the diffraction pattern of a square hole.

11.2 Plane waves of wavelength 5000 Å are incident normally on a slit of width 0.45 mm, which has a lens of focal length 900 mm following it. Find the distance from the principal maximum to the first minimum (the half-width) in the diffraction pattern formed in the focal plane of the lens.

11.3 Plane waves of wavelength 5000 Å are incident normally on a round hole of diameter 0.3 mm, which is followed by a lens of focal length 1 m. Find the distance from the principal maximum to the first minimum (the half-width) in the diffraction pattern formed in the focal plane of the lens.

11.4 If diffraction at the pupil of the eye is the limiting factor, how far away are a pair of automobile taillights resolvable? Make reasonable assumptions. Is this likely to be a realistic evaluation?

11.5 A radar set detects objects by reflecting from them radiation of wavelength 3 cm. How big should the "dish" antenna be to tell at a 10-mile distance whether or not an aircraft has its wheels down?

11.6 The single-slit diffraction pattern shown sketches one made with light of wavelength 6328 Å. The film was 2 m from the slit. How wide was the slit?

11.7 The telescope of Problem 1.6(c) in Chapter 1 has an objective 1 m in diameter. Sketch the diffraction pattern on the observer's retina for (a) a star, (b) Venus.

11.8 A single slit is illuminated by light whose wavelengths are λ_a and λ_b, so chosen that the first diffraction minimum of λ_a coincides with the second minimum of λ_b. (a) What relationship exists between the two wavelengths? (b) Do any other minima in the two patterns coincide?

11.9 A single slit is illuminated by white light. A total of seven fringes ($m = 0$, ± 1, ± 2, and ± 3) are visible. By covering the slit with red cellophane, 15 fringes are made visible. What is the "passband" of the red filter?

11.10 What is the half-width of a diffracted "beam" for a slit whose width is (a) 1, (b) 5, and (c) 10 wavelengths? The "beam" in communications jargon is the central maximum, whose full width runs from the zero at $m = -1$ to that at $m = +1$. It is given as an angular width, measured in degrees, in order to have it independent of distance. (For an antenna, the secondary maxima are known as "side lobes.")

11.11 Light whose average wavelength is λ falls on a single slit, as shown in the figure. On the screen the Nth dark fringe occurs a distance Y from the central point 0.

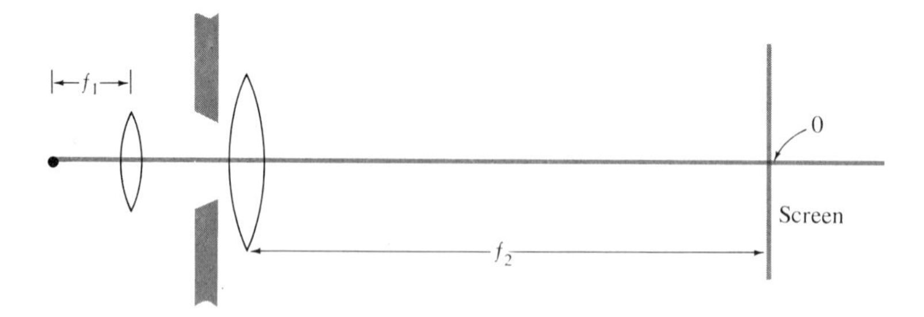

(a) How wide is the slit?

After M fringes, the pattern is too blurred to distinguish fringes.

(b) Estimate the spread of wavelengths in the source by considering just two wavelengths, λ and $\lambda + \Delta\lambda$.

(c) What is the coherence time of this source?

11.12 A single-slit diffraction pattern is seen on a screen. What changes will occur in the pattern if the entire apparatus, from light source to screen, is immersed in a medium of refractive index n?

11.13 Two small radio telescopes are substituted for one big one. The resulting fringe patterns have central maxima with the same half-width. Find the ratio of the aperture of the big telescope to the separation of the little ones.

12

Modern optics

Modern optics, broadly speaking, concerns optical effects which depend on highly coherent light, and ways of avoiding the consequences of incoherence in optical signals. Because radio waves have long coherence times, these methods are closely related to the methods of communications theory. Similarly, many of the techniques of modern nuclear and solid state physics are incorporated in the treatment of imperfectly coherent signals. In this chapter we look primarily at one of the crucial questions of modern optics — and at a very neat solution. This is the question of how to save the phase information that our long detection times generally destroy. The problem is a nice one because of its relationship to the general theory of diffraction, encountered in all the scattering techniques of modern physics, and to the problem of image formation, which we treated so differently in Chapter 1.

12.1 *Fourier transforms as diffraction patterns*

Let us consider again the problem of the single slit and ask what we have been doing *mathematically*. The geometrical construction we have worked with in phasor space is just an easy way to find the magnitude of E_T. Mathematically we have found that

$$(E_T)_I = E_1 \sin \beta_1 + E_2 \sin \beta_2 + \cdots + E_N \sin \beta_N$$

in the limit of $N \to \infty$, $E_i \to 0$, $\beta_i \to 0$. (From the I component we then infer E_T itself.) Since this procedure is the nub of the integral calculus, we might write:

$$(E_T)_I = \int_0^a E(x) \sin \beta(x)\, dx,$$

where x is the position of the source point in the slit. Of course $E(x)$ is zero except between $x = 0$ and $x = a$, in this case, so we could say

$$(E_T)_I = \int_{-\infty}^{+\infty} E(x) \sin\left(\frac{2\pi}{\lambda} x \sin \theta\right) dx$$

$$= E_0 \frac{\sin \beta/2}{\beta/2} \cdot \sin \frac{\beta}{2},$$

or

$$E_T = E_0 \frac{\sin \beta/2}{\beta/2}.$$

It is this formal procedure that must be generalized for a complete theory of diffraction. The integrals involved in such a theory are beyond the scope of this book, but the operations which they represent are quite simple. Consequently, it is worth exhibiting the general integral, without ever intending to solve it.

The reason for interest in Fraunhofer diffraction goes beyond direct applications in optics and related fields. This is because the integral of the complete theory is identified as the most common mathematical device of advanced physics, the Fourier transform. Fourier transforms change the *emphasis* of information; that is, we may have information about how particles are distributed in space, and want to emphasize their (related) distribution in momentum. We use Fourier transforms to relate these two points of view. A Fraunhofer diffraction pattern is a Fourier transform of the original object in the following sense: the phasor from some source point (x, y, z) is

$$E(x, y, z) \sin\left[\frac{2\pi}{\lambda} L - 2\pi\nu t + \phi\right],$$

where L is the distance from the source point to the observation point, (x_0, y_0, z_0), and is a function of (x, y, z). Figure 12.1 shows

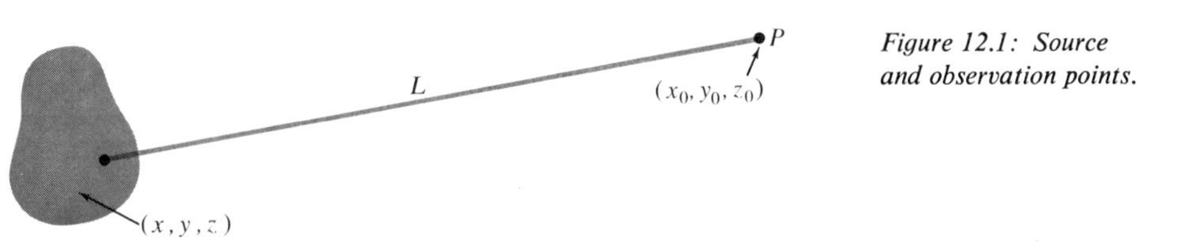

Figure 12.1: Source and observation points.

the geometry. We have already written this for the simple case of a single slit, where $y =$ constant, $z =$ constant, $L = L_0 + x \sin \theta$.

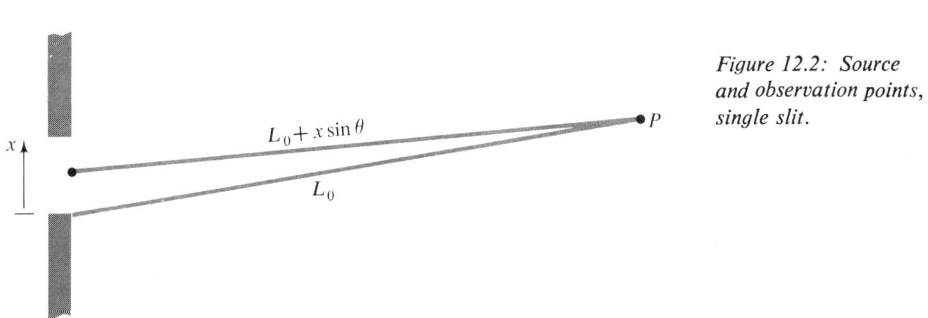

Figure 12.2: Source and observation points, single slit.

In the general case, the amplitude of the electric field at P is the "sum" of many phasors of the type we have written:

$$E_T(x_0, y_0, z_0) = \iiint E(x, y, z) \sin\left[\frac{2\pi L}{\lambda} - 2\pi\nu t + \phi\right] dx\, dy\, dz.$$

Mathematically, this is a *Fourier transform*, and we say that $E_T(x_0, y_0, z_0)$ is the function which is the transform of $E(x, y, z)$. The two functions are completely unalike, yet both contain the same information, differently organized.

If E_T is the Fourier transform of E, then E is said to be the inverse transform of E_T. An important aspect of Fourier transforms is that E is also the Fourier transform of E_T, as shown in Appendix H. The function we are using here is called the Fourier sine transform, which is further generalized by the use of complex exponentials, in the usual form employed in physics.

Another example of the reorganization of information upon Fourier transformation is the "measurement" of spectral line shapes by an interferometer. In this case we relate the intensity distribution as a function of wavelength to that with respect to plate separation at $\theta = 0$:

$$E_{\mathrm{T}}(d) = \int E\left(\frac{1}{\lambda}\right) \cos\left(\frac{2\pi d}{\lambda}\right) d\left(\frac{1}{\lambda}\right).$$

This is the most common use of the Michelson interferometer and is most important to infrared work, where gratings and prisms are difficult. The application is essentially that discussed in Chapter 10.

Gratings also sort out the Fourier components of a light source when they exhibit the spectrum. Fourier transforms occur repeatedly throughout physics and mathematics, but nowhere so dramatically as in Fraunhofer diffraction. In fact, it is possible to use diffraction to perform transformations which would be tedious or impossible to calculate.

12.2 *Image retrieval*

Now we might ask: how is it that we are not always having to contend with diffraction patterns in optical devices? The answer, of course, is that with lenses we are able to retrieve an image that has the same

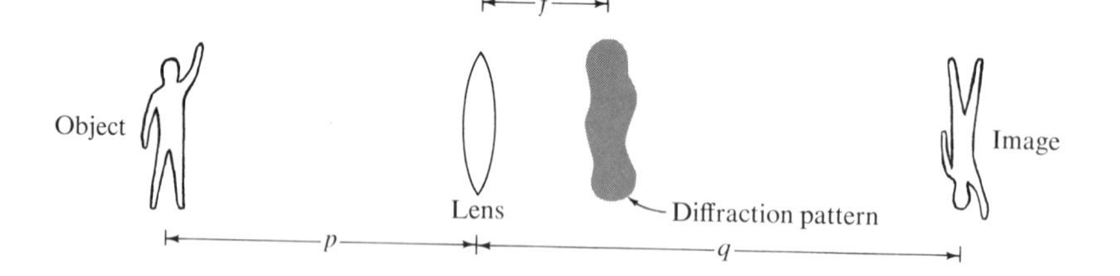

Figure 12.3: Diffraction pattern and image formation.

shape as the original object. In Figure 12.3, the lens not only brings the diffraction pattern in from infinity, but also retransforms (or "untransforms") it into an image. Now the Fourier transformation just reorganizes information, so the lens is simply organizing it back again. This means that if we take some of the information out of the diffraction pattern, a different message gets retransformed and the image will be different. This is easy to demonstrate in the following experiment*:

We use as an object one whose diffraction pattern is simple, a lattice formed by crossing two coarse gratings. By masking off

* This was first devised by A. B. Porter, *Phil. Mag.*, **B11**, 154 (1906). *See also* Dutton, Givens, and Hopkins, *Am. J. Phys.*, **32**, 355 (1964).

the diffraction pattern, we make the uninterrupted part identical to the diffraction pattern of a different object. Figure 12.4 sketches two

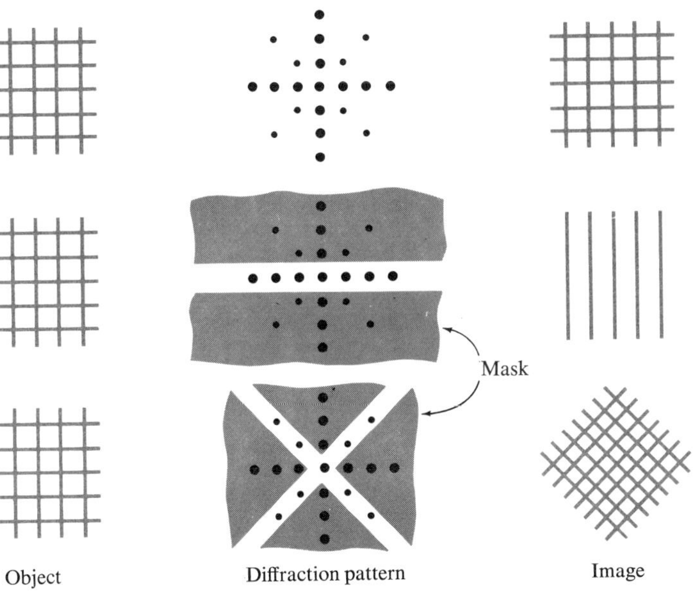

Figure 12.4: Porter's experiment.

simple examples. In each case the image obtained is of the object which would give a diffraction pattern like the one left after the masking. An exactly analogous experiment (shown in Figure 12.5) would be to construct a monochromatic image with a prism and masks. The prism would disperse the light from an object, so that we would see colored images. The mask then would select one of these Fourier components. A second prism would then "untransform" the information, although this is not really necessary because, unlike a diffraction pattern, a colored image is still intelligible.

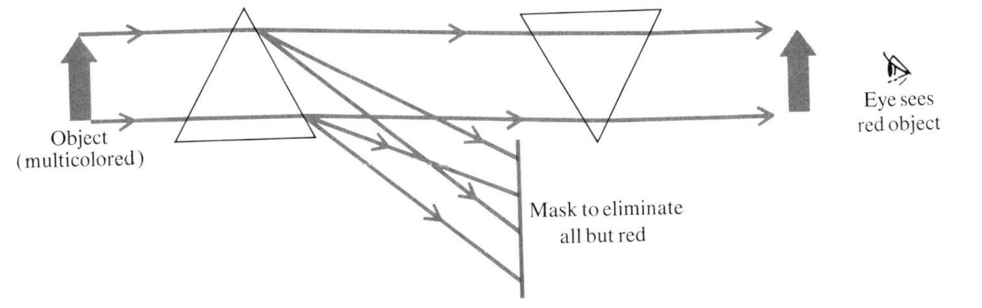

Figure 12.5: Fourier analysis with a prism.

When we use x-rays we have no lens. Consequently, image formation is not possible, and information about the shape of an object is left in a diffraction pattern. The information is still there, but it is not in a very useful form unless the diffracting object is of simple enough shape so that its pattern is familiar. Thus, single crystals, in which the atoms form a regular array, produce patterns like those of our crossed gratings (the only difference being that the crystals are three-dimensional). Since knowledge of the positions of atoms in materials is worth a fair amount of effort, x-ray diffraction is a useful technique. Its major limitation is that one must "guess" an object, calculate its diffraction pattern, and compare. (Also, only the heavy, many-electron atoms scatter a sufficient intensity of x-rays to be "seen".)

It is not, in fact, possible to retransform mathematically because our detecting devices (usually photographic film) record I_T rather than $\mathbf{E}_T$ so that we lose the necessary information on phase. If we had detectors sufficiently rapid, or some means of preserving the phase information, we might expect to form a true image of even complicated objects like biological molecules. Such a technique has recently become available for visible light, through the invention of the laser and the accompanying realization of holography.

12.3 *Holography—recovery of phase information*

Holography is a technique for preserving the phase information needed to retransform a diffraction pattern without a lens. As an example, remember the pattern of the crossed gratings: in this case each bright spot has the same phase so that we have almost enough information to use a photograph of the diffraction pattern as an object and recover the image of crossed gratings as its diffraction pattern (remember that the Fourier transform of a Fourier transform is the original). One reason for saying "almost" here is that, according to Babinet's principle, we might find that we had an image of the complementary object. Another reason is that only the centers of the spots really have the same phases.

To resolve the ambiguity due to Babinet's principle, all we need to know is which of two possible phases the light had when it exposed the film. How can we record this information? One way is to expose the film simultaneously to a wave that is coherent with the waves forming the diffraction pattern. The phase of this reference wave varies simply across the film, and the diffracted wave interferes with it, since they are coherent. Thus the film forms a record of the phase of the diffraction pattern relative to the reference beam. The phase information is preserved because the film records

$$(E_{\text{sig}} + E_{\text{ref}})^2 = E_s^2 + E_r^2 + 2E_s E_r,$$

and if $E_r \gg E_s$, this is approximately $E_r^2 + 2E_s E_r$, where E_r^2 is just a uniform background. So the intensity is proportional to E_s instead of E_s^2, and we can tell whether $E_s = +|E_s|$ or $-|E_s|$. This is illustrated

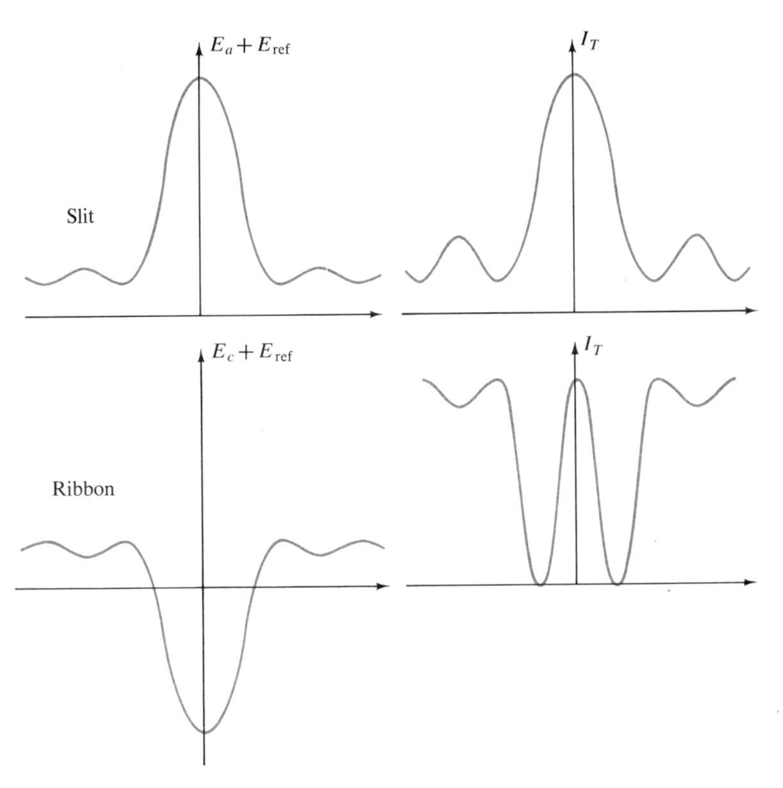

Figure 12.6: Decoding the Babinet pair.

in Figure 12.6. The technique is similar to well-known methods in communications.

Next we let a plane wave be diffracted by the pattern on the film, to form a Fourier transform of the new $I_T (\propto E_s)$, and hence a true image of the original object. This image has lost none of the depth information from the original, since the diffraction patterns of objects at different distances to the film have different phases relative to the reference beam, and this information is recorded. Such a situation corresponds to infinite "depth of focus", and the image is truely three-dimensional. The whole discussion, of course, is applicable to objects and diffraction patterns of much greater complexity than the simple one we have treated. We will consider an actual holographic camera in order to explore the technique in more detail. For this we use an equally valid, but quite different, approach to the phenomenon.

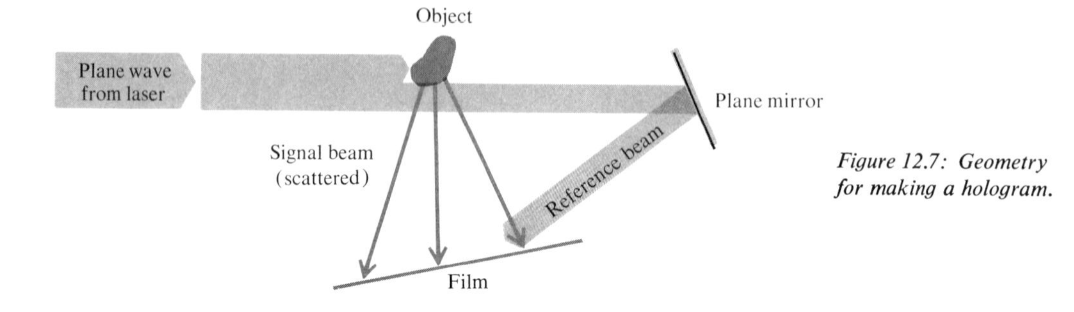

Figure 12.7: Geometry for making a hologram.

12.4 *Holography — "gratings" approach*

Figure 12.7 illustrates the arrangement of the apparatus. This is a simple way of deriving two beams that are coherent to each other, and corresponds to the two-slit (spatial separation) way of achieving simple interference. First think of an object which is a second plane mirror, as in Figure 12.8, so that both signal and reference beam are plane

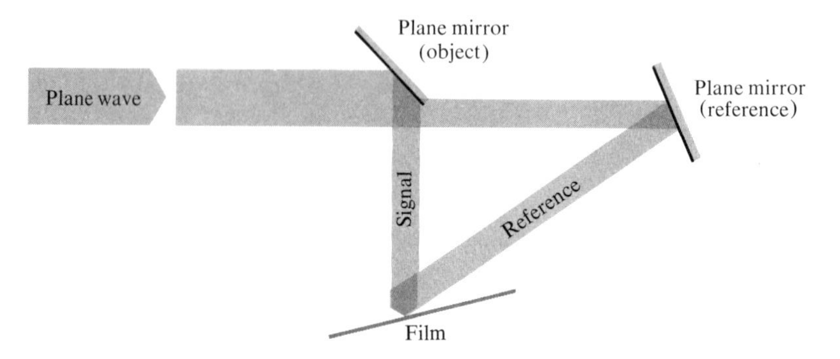

waves, making some angle α with each other (remember Problem 8.4, Chapter 8). Either beam alone would blacken the film uniformly, but together they produce interference fringes. The film is therefore darkened in lines, spaced $(\lambda/\sin\alpha)$ apart. This pattern on the exposed film is a grating, whose diffraction pattern in turn is the familiar set of spots at the angles $\sin\theta = m\lambda/d$. But we know d, so the spots turn out to occur at $\sin\theta = m\lambda/(\lambda/\sin\alpha) = m\sin\alpha$. If the film has sufficiently high resolution (that is, if it can record fine details of the pattern), the dark lines give way to the bright ones as a sinusoidal function rather than abruptly, with the result that only orders $m = 0, \pm 1$ occur. This diffraction pattern is a holographic reconstruction of the two infinitely distant point sources we would see if we looked at the original beams from the position of the film. It is observed by looking through the film at a point source. The zeroth order spot (the image of the reference beam) will be present in all our reconstructions. Two images of the signal beam occur, corresponding to the orders $m = \pm 1$.

Next replace the object mirror with two mirrors, as shown in Figure 12.9. The diffraction (interference) pattern is more complicated;

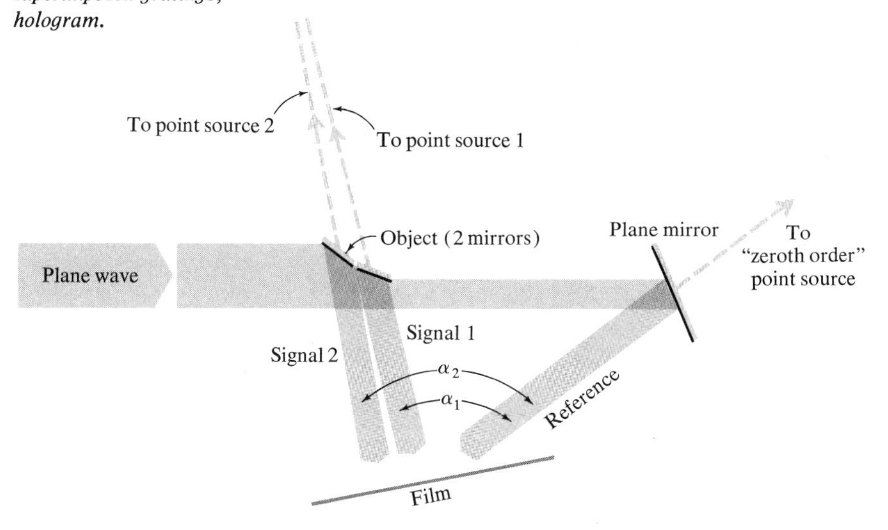

Figure 12.9: Two superimposed gratings, hologram.

there are two superposed sets of "gratings". Each in turn yields spots as its reconstruction. Obviously, this process can be carried on to ever more complicated arrangements of mirrors and to as complex patterns of reconstructed spots as desired.

Rather, let us next replace our object mirrors with two scattering centers, as in Figure 12.10. The waves which now impinge on the film

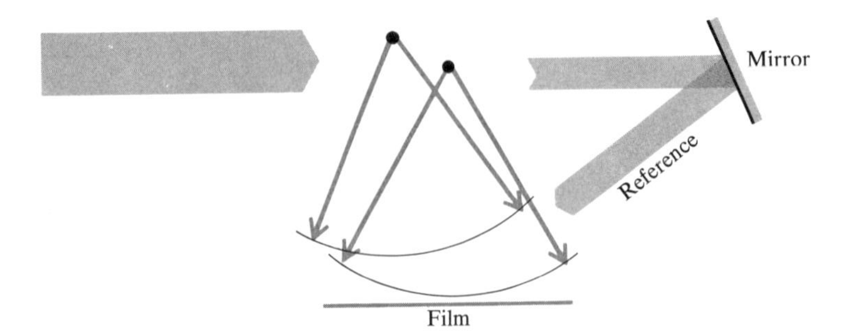

Figure 12.10: Hologram of two scattering centers.

form "gratings" whose spacing varies because the angle between scattered and reference beams changes across the film. Similarly, the reconstruction shows scattering spots at different angles for different parts of the film, which means that they have perspective and can be perceived as sources close to the film rather than at infinity. Now multiply our spots arbitrarily, and a complicated, diffusely scattering object can be photographed in this way.

Upon looking through the film at a distant point source, as indicated in Figure 12.11, we see the original object as a complete three-

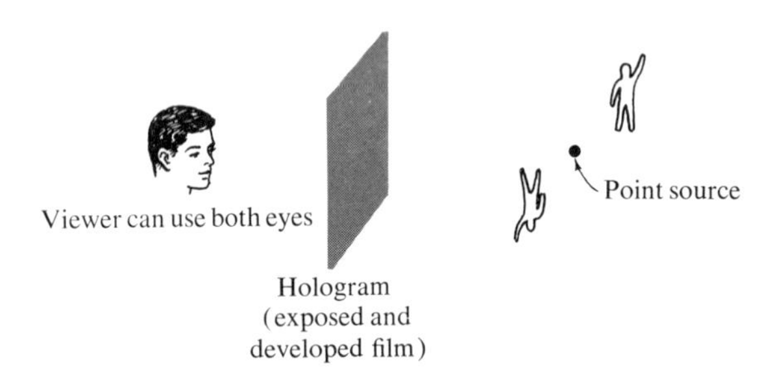

Figure 12.11: Viewing the holographic reconstruction.

dimensional reconstruction. As in the simple case, a second ($m = -1$) image appears, in focus at a different point. If the angle α is large enough, the $m = 0$ and $m = -1$ images are out of the "picture".

All of the effects due to diffraction can be studied by themselves, but holograms illustrate them completely and dramatically. The phenomena of modern optics are those which have been known (sometimes only theoretically) before, but are now accessible to investigation because the laser provides an intense coherent light source.

PROBLEMS

12.1 A hologram film shrinks 20 percent when processed. What is the resulting change in the reconstruction? How might this be compensated?

12.2 In the experiment on the imaging of crossed gratings, what image will appear if we block off every other line of diffraction spots?

12.3 For the diffraction pattern of a square hole devise a "mask" which will result in a slit being imaged.

12.4 A film used in holography is capable of recording a pattern of lines 1 μm apart. What is the maximum possible angle between the two beams if one of them is normal to the film? Use light of wavelength 6328Å. This limits the closeness of an object to the film. If the reference beam strikes the film normally, how large an area of the film will be useful if the object is 3 cm from the film?

Appendix A: Miscellaneous mathematical notes

A.1 $\sin A = \cos(\frac{1}{2}\pi - A) = -\sin(-A) = -\sin(A + \pi).$

$$\cos A = \sin(\tfrac{1}{2}\pi - A) = +\cos(-A) = -\cos(A + \pi).$$

$$\sin A + \sin B = 2 \sin\left(\frac{A+B}{2}\right) \cos\left(\frac{A-B}{2}\right).$$

$$\sin(\alpha + \beta) = \sin \alpha \cos \beta + \sin \beta \cos \alpha.$$

A.2 If δ is a small number, a useful approximation is $(1 + \delta)^s \cong 1 + s\delta$. This comes from the binomial series:

$$(1 + \delta)^s = 1 + s\delta + \frac{s(s-1)}{2}\,\delta^2 + \cdots \qquad \delta^2 < 1.$$

Note that both s and δ may be negative, and s need not be an integer.

A.3 Let $E(t) = E_0 \sin(\alpha + 2\pi/\tau)$.

$$\langle E(t)\rangle_{av} = \frac{1}{T}\int_0^T E(t)\,dt = \frac{1}{T}E_0\frac{\tau}{2\pi}\int_{t=0}^{t=T}\sin\left(\alpha+\frac{2\pi t}{\tau}\right)d\left(\alpha+\frac{2\pi t}{\tau}\right)$$

$$= \frac{E_0}{2\pi}\left(\frac{\tau}{T}\right)\left\{-\cos\left(\alpha+\frac{2\pi T}{\tau}\right)+\cos\alpha\right\}.$$

The quantity in { } is never greater than ± 2, no matter what value T has. But the whole thing is proportional to τ/T. Therefore, the average is very small if $T \gg \tau$. We only need look at the term in { } if this is not the case. So $\langle E(t)\rangle_{av} \simeq 0$ for $T \gg \tau$.

A.4 This is not the case with $E^2(t)$:

$$E^2(t)_{av} = \frac{1}{T}\int E^2(t)\,dt$$

$$= \frac{E_0{}^2}{2\pi}\frac{\tau}{T}\int_{t=0}^{t=T}\sin^2\left(\alpha+\frac{2\pi t}{\tau}\right)d\left(\alpha+\frac{2\pi t}{\tau}\right)$$

$$= \frac{E_0{}^2}{2\pi}\frac{\tau}{T}\left\{\frac{1}{2}\left(\alpha+\frac{2\pi t}{\tau}\right)-\frac{1}{4}\sin 2\left(\alpha+\frac{2\pi t}{\tau}\right)\Bigg|_0^T\right\}$$

$$= \frac{E_0{}^2}{2}\frac{\tau}{T}\left\{\frac{\pi T}{\tau}-\frac{1}{2}\sin 2\left(\alpha+\frac{2\pi T}{\tau}\right)+\frac{1}{4}\sin 2\alpha\right\}$$

$$= \frac{1}{2}E_0{}^2 + \frac{E_0{}^2}{8\pi}\frac{\tau}{T}\left\{\sin 2\alpha - \sin 2\left(\alpha+\frac{2\pi T}{\tau}\right)\right\}.$$

Here, for all values of T, there is a constant term of $\frac{1}{2}E_0{}^2$. The second term behaves similarly to that in $\langle E(t)\rangle_{av}$, and is small when $T \gg \tau$. So

$$\langle E^2(t)\rangle_{av} = \frac{1}{2}E_0{}^2 \qquad \text{when } T \gg \tau.$$

In the case of a spatial average such as $\langle E(x)\rangle_{av}$, the manipulations and conclusions are the same, with λ playing the same role as τ plays here.

A.5 *Partial Derivatives.* We are familiar with the derivative:

$$\frac{df(x)}{dx} = \lim_{\Delta x\to 0}\frac{f(x+\Delta x)-f(x)}{\Delta x}.$$

Similarly, we can define a *partial* derivative:

$$\frac{\partial f(x, y)}{\partial x} = \lim_{\Delta x \to 0} \frac{f(x + \Delta x, y) - f(x, y)}{\Delta x}$$

where $f(x, y)$ is treated *as if* it were a function of x only, with y held constant. Then we find a general expression:

$$\frac{df(x, y, z)}{dz} = \frac{\partial f}{\partial x}\frac{dx}{dz} + \frac{\partial f}{\partial y}\frac{dy}{dz} + \frac{\partial f}{\partial z}.$$

A.6 *Vectors.* The vector $\mathbf{V}$ is a quantity having the magnitude V and the direction $\mathbf{V}/V = \hat{\mathbf{v}}$. We call $\hat{\mathbf{v}}$ a *unit* vector. It has magnitude 1 and direction along $\mathbf{V}$. So $\mathbf{V} = V\hat{\mathbf{v}}$. This is useful when we want to take components in, say, a Cartesian system $\mathbf{V} = \hat{\mathbf{i}}V_x + \hat{\mathbf{j}}V_y + \hat{\mathbf{k}}V_z$, where $\hat{\mathbf{i}}, \hat{\mathbf{j}}, \hat{\mathbf{k}}$ are unit vectors along the x, y, z axes. Note that

$$\hat{\mathbf{i}} \cdot \mathbf{V} = V_x \quad \text{and} \quad \hat{\mathbf{i}} \times \hat{\mathbf{j}} = \hat{\mathbf{k}}.$$

A.7 *Units.*

1 millimeter (mm)	10^{-3} m	1 picometer (pm)	10^{-12} m
1 micrometer (μm) [written 1μ (*micron*) until recently]	10^{-6} m	1 hertz (Hz)	1 cps
		1 kilohertz (kHz)	10^3 Hz
1 nanometer (nm)	10^{-9} m	1 megahertz (MHz)	10^6 Hz
1 angstrom (Å)	10^{-10} m	1 gigahertz (GHz)	10^9 GHz

Appendix B: Derivation of the wave equation

B.1
Wave equation — string

The wave equation tells us, in effect, why the wave propagates. Consider the triangular pulse shown in Figure B-1 as it moves along the string. The tension ahead of the pulse and behind it is T, and this

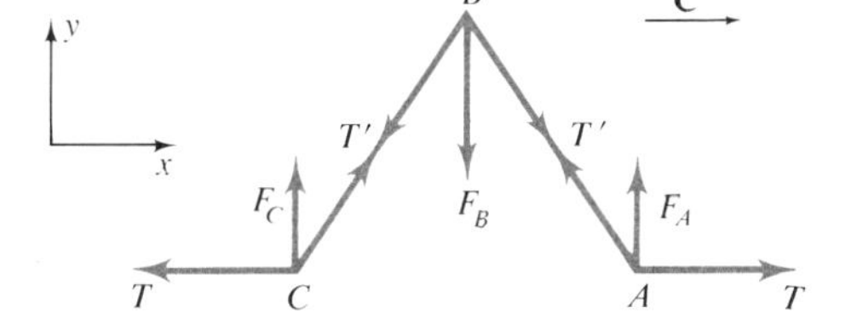

Figure B-1

force is in balance so that the string does not move. That is, at each point the forces look like Figure B-2.

Figure B-2

The same is true on the sides of the pulse, where the string is straight. The string is stretched here, so the tension is a different one,

T'. Only at the "corners" are there unbalanced forces. For instance, at point A there is a net force F_A upward, and we expect that little piece of string to be accelerated upward.

Between A and B each piece of string is in uniform motion (upward), since there are no unbalanced forces. At B a force stops the motion and reverses it. And so the pulse progresses down the string. The critical places are those where the slope of the string changes. We will generalize this as follows: The small segment of string shown in Figure B-3 is acted on by the tensions $\mathbf{T}_1$ and $\mathbf{T}_2$. Its

Figure B-3

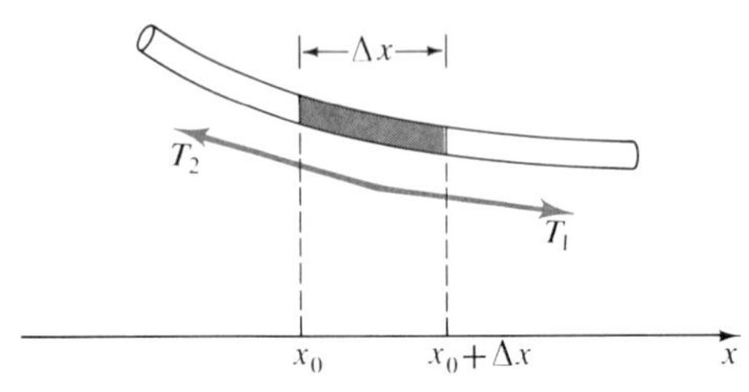

length, when in its equilibrium position, is Δx. The linear density of the string is ρ, so the mass of the segment is $\rho\ \Delta x$.

The segment does not move in the x direction, so the x components for the forces $\mathbf{T}_1$ and $\mathbf{T}_2$ are equal: $(T_1)_x = (T_2)_x = T$. We apply Newton's second law to the unbalanced forces in the y direction:

$$-(T_1)_y + (T_2)_y = \rho\ \Delta x \cdot \frac{\partial^2 y}{\partial t^2}$$

(taking positive forces along $+y$ and inserting the signs explicitly).

Figure B-4

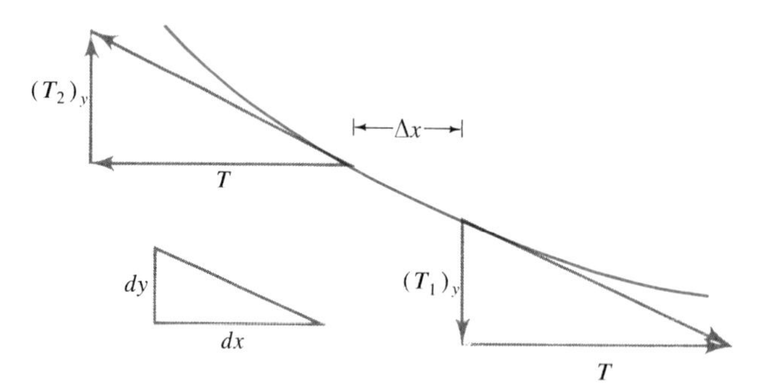

Now, by geometry (Figure B-4), we see that the ratio

$$\left(\frac{\partial y}{\partial x}\right)_2 = -\frac{(T_2)_y}{T}, \qquad \text{and} \qquad \left(\frac{\partial y}{\partial x}\right)_1 = -\frac{(T_1)_y}{T}.$$

Thus

$$\frac{\rho\,\Delta x}{T}\frac{\partial^2 y}{\partial t^2} = +\left(\frac{\partial y}{\partial x}\right)_1 - \left(\frac{\partial y}{\partial x}\right)_2$$

or

$$\frac{\rho}{T}\frac{\partial^2 y}{\partial t^2} = \frac{\left(\frac{\partial y}{\partial x}\right)_{\text{at }(x_0+\Delta x)} - \left(\frac{\partial y}{\partial x}\right)_{\text{at }x_0}}{\Delta x}$$

According to our standard definition of a derivative, this becomes, in the limit as $\Delta x \to 0$,

$$\frac{\rho}{T}\frac{\partial^2 y}{\partial t^2} = \frac{\partial^2 y}{\partial x^2}$$

This is the wave equation for the string. We see, then, that $c^2 = \rho/T$, so that the phase velocity (wave speed) depends on the physical properties of the string.

B.2 *Sound* Sound waves differ from waves on a string or on the surface of water in that the motion of the medium is *along* the direction of travel of the wave rather than *transverse* to it. We wish now to derive the wave equation that governs the pressure fluctuations, which we call "sound", in a gas. Each molecule moves back and forth, but we assume none ever travels as much as a wavelength, as shown in Figure

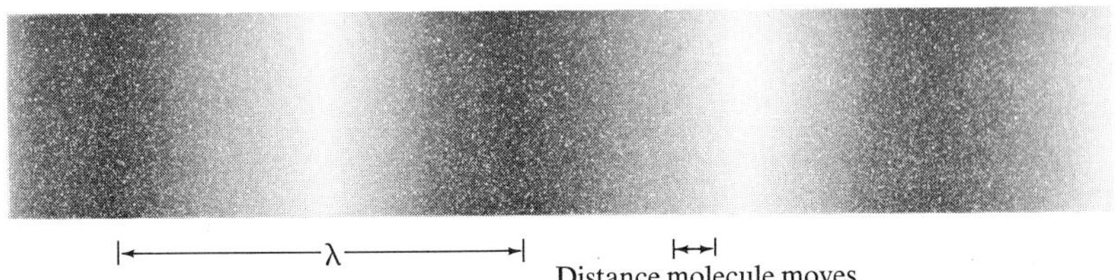

Figure B-5

Although it is the pressure which we detect as changing, the density is a related function and could also be used to characterize the wave. So $p = p(x, t)$ and $\rho = \rho(x, t)$. The values of pressure and density, however, change very little compared with their equilibrium values, so we might say $p(x, t) = p_0 + \Delta p(x, t)$, where p_0 is a large constant and Δp is always small in comparison with it; similarly, $\rho(x, t) = \rho_0 + \Delta\rho(x, t)$.

Now let us follow a given mass of air molecules as they move to produce the sound wave: we look at a cylinder of cross-sectional area A. Since the molecules move only in the $\pm x$ direction, the cylinder retains its area, and only its length changes. At equilibrium, the gas molecules fill the volume shown by the dotted lines in Figure B-6. When displaced, the whole volume both moves and ex-

Figure B-6

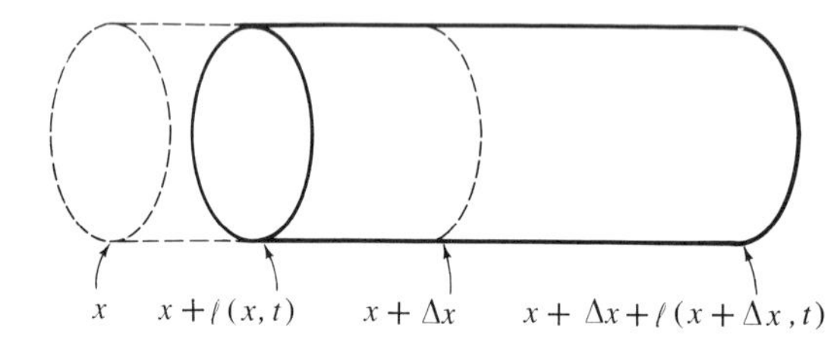

pands (or contracts). This means that the molecule on the left end may move a distance $\ell(x, t)$ from the equilibrium position x, but a molecule from the right end (equilibrium position $x + \Delta x$) moves a different distance, $\ell(x + \Delta x, t)$. Then

Equilibrium volume $= A \cdot \Delta x = V_{\text{eq}}$.

Displaced volume $= A\{\Delta x + \ell(x + \Delta x, t) - \ell(x, t)\} = V_{\text{dis}}$

Now the mass contained in each volume is the same, so we can say that

$$M = \rho_0 V_{\text{eq}} = \rho V_{\text{dis}},$$

$$\rho_0 A \, \Delta x = (\rho_0 + \Delta\rho)A\{\Delta x + \ell(x + \Delta x, t) - \ell(x, t)\},$$

whence

$$\Delta\rho = -\rho_0 \frac{\ell(x + \Delta x, t) - \ell(x, t)}{\Delta x}.$$

In the limit of vanishingly small Δx (which is the only sensible volume to talk about),

$$\Delta\rho = -\rho_0 \frac{\partial \ell}{\partial x}. \tag{1}$$

Now, we may not know how p and ρ are related, but we can say that p is a function of ρ (for example, $p = \text{const} \cdot \rho^{\gamma}$ is the equation for a perfect gas with no heat flowing in or out, an adiabatic). So, consider $p = f(\rho)$:

$$p_0 = f(\rho_0), \qquad \text{so } p_0 + \Delta p = f(\rho_0 + \Delta \rho).$$

Subtracting,

$$\Delta p = \Delta\rho \frac{f(\rho_0 + \Delta\rho) - f(\rho_0)}{\Delta\rho}.$$

Then, in the limit as $\Delta\rho \to 0$, this can be written as

$$\Delta p = \Delta\rho\left(\frac{\partial f}{\partial \rho}\right)_0 = \Delta\rho\left(\frac{\partial p}{\partial \rho}\right)_0. \tag{2}$$

Now we are ready to insert the physics: We write Newton's second law for the molecules in our volume:

$$F_{\text{net}} = Ma \qquad \text{or} \qquad p_{\text{net}} \cdot A = (\rho_0 \cdot A\, \Delta x)\frac{\partial^2 x}{\partial t^2}.$$

The position of a molecule is given by x, but we can write $x = (x_{\text{eq}} + \ell)$, and only ℓ changes; then

$$\frac{\partial^2 x}{\partial t^2} \equiv \frac{\partial^2 \ell}{\partial t^2},$$

and:

$$\begin{aligned} p_{\text{net}} &= [p_0 + \Delta p \text{ (at left end)}] - [p_0 + \Delta p \text{ (at right end)}] \\ &= \Delta p(x, t) - \Delta p(x + \Delta x, t). \end{aligned}$$

So:

$$-\frac{\Delta p(x + \Delta x, t) - \Delta p(x, t)}{\Delta x} = \rho_0 \frac{\partial^2 \ell}{\partial t^2}.$$

Again we let $\Delta x \to 0$; so,

$$\rho_0 \frac{\partial^2 \ell}{\partial t^2} = -\frac{\partial}{\partial x}\Delta p. \tag{3}$$

Now use Eq. (2):

$$\rho_0 \frac{\partial^2 \ell}{\partial t^2} = -\frac{\partial}{\partial x}\Delta\rho\left(\frac{\partial p}{\partial \rho}\right)_0 = -\left(\frac{\partial p}{\partial \rho}\right)_0 \frac{\partial\, \Delta\rho}{\partial x}$$

and Eq. (1)

$$= -\left(\frac{\partial p}{\partial \rho}\right)_0 (-\rho_0) \frac{\partial}{\partial x} \frac{\partial \ell}{\partial x}.$$

So

$$\frac{\partial^2 \ell}{\partial t^2} = \left(\frac{\partial p}{\partial \rho}\right)_0 \frac{\partial^2 \ell}{\partial x^2}. \tag{4}$$

This is a wave equation on ℓ, with solutions

$$\ell = \ell(x + ct) \qquad \text{where } c = \sqrt{\left(\frac{\partial p}{\partial \rho}\right)_0}.$$

We can again use Eqs. (1) and (2) to show that if

$$\ell = \ell_0 \sin \frac{2\pi(x - ct)}{\lambda},$$

then

$$\Delta p = P \cos \frac{2\pi(x - ct)}{\lambda},$$

where $P = -2\pi\rho_0 c^2/\lambda$.

An interesting physical consequence of this calculation is that we expect the speed of sound to be $c = \sqrt{(\partial p/\partial \rho)_0}$, which is $c = \sqrt{\gamma kT/m}$ for a perfect gas, a fact which allows us to measure γ.

Appendix C: Strings

C.1 *Two dissimilar strings*

We assume that initially a wave $y_i(x - c_1 t)$ is moving on string (1) toward the boundary at x_B, as shown in Figure C-1. After the wave reaches the boundary, a wave is transmitted onto the second string.

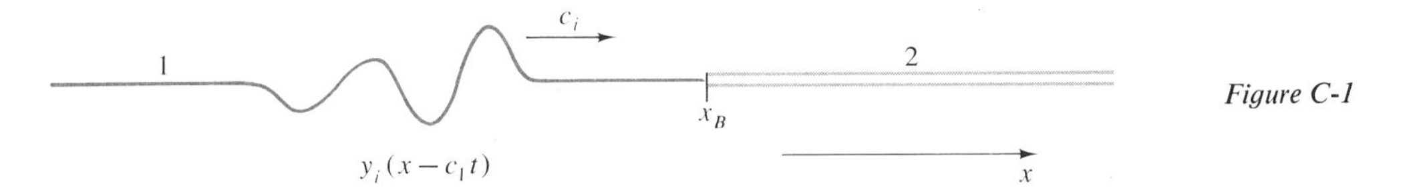

Figure C-1

There the wave has the same shape, but may be smaller, and moves at a different speed. At the same time, another wave of the same shape may be generated at the boundary and move back along the first string. These transmitted and reflected waves are described by $y_t(x - c_2 t)$ and $y_r(x + c_1 t)$.

The relationships among these waves are found from the boundary conditions:

1. At the boundary, the string is continuous so that the net

displacement is the same whether we look just to the right or just to the left of the boundary:

$$y_i(x_B - c_1 t) + y_r(x_B + c_1 t) = y_t(x_B - c_2 t).$$

This is true for all times, a convenient one being $t = 0$. We use the fact that all the waves have similar shapes as follows: let $y_i = A_i f(x - c_1 t)$, where A_i is the amplitude $y_i(x_B, 0) = A_i$. This implies a choice of units such that $f(x_B, 0) = 1$.

Similarly, $y_t = A_t f(x - c_2 t)$ and $y_r = A_r f(x + c_1 t)$, where f is the same function for all cases. Then our boundary equation is

$$A_i + A_r = A_t, \qquad \text{at } t = 0.$$

These amplitudes are constants, so the statement is true for all t.

2. The second boundary condition makes use of the fact that the boundary point has a single velocity, v_y. Thus,

$$\left(\frac{\partial y_i}{\partial t}\right)_{x_B} + \left(\frac{\partial y_r}{\partial t}\right)_{x_B} = \left(\frac{\partial y_t}{\partial t}\right)_{x_B}.$$

Now

$$\frac{\partial y_i}{\partial t} = -c_1 A_i \frac{\partial f(x - c_1 t)}{\partial (x - c_1 t)} = -c_1 A_i g(x - c_1 t)$$

and

$$\frac{\partial y_t}{\partial t} = -c_2 A_t \frac{\partial f(x - c_2 t)}{\partial (x - c_2 t)} = -c_2 A_t g(x - c_2 t).$$

The point here is that the functions g, like the functions f above, are the same functions, though of different argument. Thus at $t = 0$ and $x = x_B$,

$$-c_1 A_i + c_1 A_r = -c_2 A_t.$$

Again, these are constants, so the equation is true for all values of t. Combining the two boundary statements, we get

$$A_r = A_i \frac{c_1 - c_2}{c_1 + c_2}, \qquad A_t = A_i \frac{2c_1}{c_1 + c_2}.$$

First we should check these with the special cases we already know:

(a) $c_1 = c_2$: the two strings have the same phase velocity (wave speed):

$A_t = A_i, \qquad A_r = 0,$

so the wave passes x_B unchanged.

(b) $c_2 \gg c_1$: the second string is rigid or very massive, so the boundary is effectively fixed: $A_t = 0$, $A_r = -A_i$. This means that there is no transmitted wave and the reflected one is simply the incident one inverted.

(c) $c_1 \gg c_2$: the second string is so loose or light that the boundary may be considered free:

$A_t = 2A_i, \qquad A_r = +A_i.$

Here the reflected wave is not inverted. The transmitted wave is large, but moves very slowly.

C.2 *Mechanics of reflection of a pulse*

Chapter 3 describes reflection mathematically, but it should be understandable in physical terms.

In order to understand this as thoroughly as we should, let us follow a simple pulse and apply Newton's laws as we go. To start with, we must state these laws carefully. The first tells us that if there is no net force on a particle, the particle is not accelerated, but continues to move at a constant velocity (often zero). The second law says that if a net force does exist on the particle, it is accelerated according to the equation $\mathbf{F}_\mathrm{T} = m\mathbf{a}$, where $\mathbf{F}_\mathrm{T}$ is the *net* force left unbalanced, m is the particle mass, and $\mathbf{a}$ is the acceleration.

The third law is the difficult one. This states that if one particle, A, exerts a force $\mathbf{F}_{AB}$ on B, then particle B will exert an equal and opposite force, $\mathbf{F}_{BA} = -\mathbf{F}_{AB}$, on A. The point to remember is that $\mathbf{F}_{AB}$ acts *on* B, while $\mathbf{F}_{BA}$ acts *on* A, and therefore both particles may be accelerated by these forces. It may be that F_{AB} is balanced out, but it cannot be balanced by F_{BA}, which acts on a different particle.

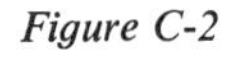

Figure C-2

Figure C-2 shows a very simple pulse on a string made of beads

strung on a massless rubber band. Force may be exerted on a bead only by the tension in the band. Since the tension changes only at beads, we can say that the forces are exerted by the neighboring beads. Thus bead b is acted on by two forces only, as shown in Figure C-3.

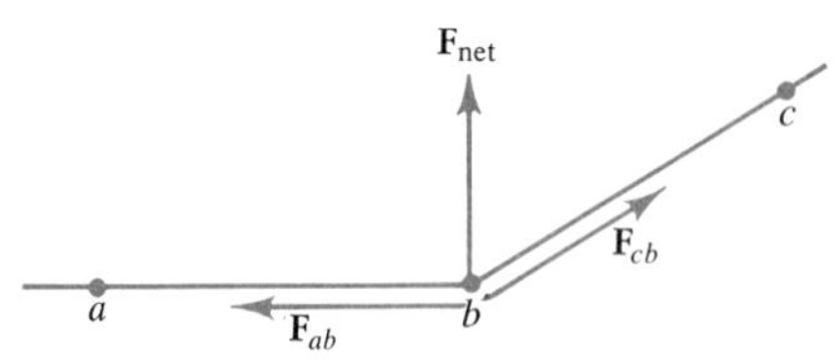

Figure C-3. $\mathbf{F}_{ab}$ pulls toward bead a, and $\mathbf{F}_{cb}$ pulls toward bead c. Their horizontal components balance and leave a net force upward so that b is accelerated. As far as bead b is concerned, we need not worry about reaction (third law) forces to $\mathbf{F}_{ab}$ and $\mathbf{F}_{cb}$, since these act on beads a and c. For instance, $\mathbf{F}_{ba} = -\mathbf{F}_{ab}$ is a force that acts on bead a and may in turn be balanced by some force acting on a from its other side.

Since all horizontal components of force are balanced, we show only the vertical components in Figure C-4, which is a sequence of

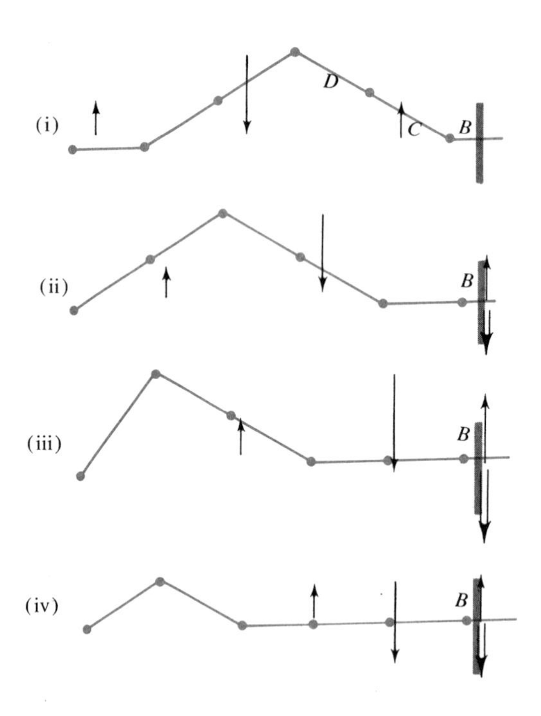

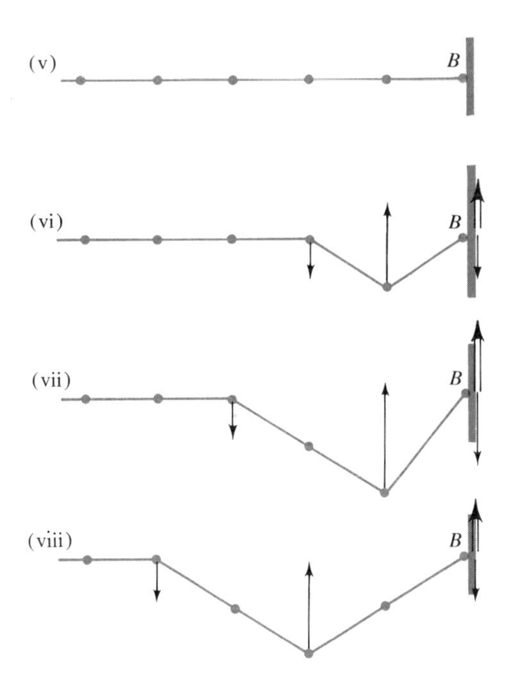

snapshots of the pulse. The single arrows indicate *net* forces on beads due to other beads. Only B is acted on by the wall (double arrow). All forces acting on B must sum to zero, since B does not move. Thus we know that in Figure C-4(ii), the force exerted on B by the wall is exactly enough to balance that exerted by C. Here the forces on C due to D and B are just balanced, and it moves uniformly without acceleration. In Figure C-4(iii), the forces on B are larger, but still total zero. The force on C is now unbalanced, and is somewhat bigger than that exerted previously on D and E. Unlike the other beads, C experiences no decelerating forces until it has passed its equilibrium position (again because B cannot move up). Thus it overshoots until it reaches the configuration of Figure C-4(vii), when the new (inverted) pulse has been launched.

Appendix D: Summary on electricity and magnetism — derivation of electromagnetic wave equation

The *force* exerted on a charge of magnitude q, traveling with velocity $\mathbf{v}$ is

$$\mathbf{F} = q[\mathbf{E} + \mathbf{v} \times \mathbf{B}],$$

which may be taken as a definition of the electric field $\mathbf{E}$ and the magnetic induction $\mathbf{B}$. Note that $\mathbf{E}$ can do work, since the charge may move parallel to the force, but $\mathbf{B}$ cannot, since $\mathbf{v} \times \mathbf{B}$ is always perpendicular to $\mathbf{v}$ and $\mathbf{B}$.

The *energy* stored in the fields is

$$U = \mathscr{E} \cdot \text{volume} = \frac{1}{2}\epsilon \iiint_V E^2\, dV + \frac{1}{2}\frac{1}{\mu} \iiint_V B^2\, dV.$$

Here the $\iiint_V dV$ means an integration over volume. Thus the energy *density* is

$$\mathscr{E} = \frac{U}{V} = \frac{\epsilon}{2} E^2 + \frac{1}{2\mu} B^2.$$

The parameters ϵ and μ characterize the medium in which the fields exist. We will find that the index of refraction is $n = \sqrt{\epsilon/\epsilon_0}$, where ϵ_0 is ϵ for vacuum.

Maxwell's equations are statements about the sources of the fields:

$$\oiint \epsilon \mathbf{E} \cdot d\mathbf{S} = Q. \tag{1}$$

(Gauss' law: a source of **E** is charge. Q is all the charge inside the closed surface.)

$$\oiint \mathbf{B} \cdot d\mathbf{S} = 0. \tag{2}$$

(The same, only that there is no magnetic charge.)

$$\oint_C \mathbf{E} \cdot d\mathbf{l} = -\frac{d}{dt} \iint_\Sigma \mathbf{B} \cdot d\mathbf{S} \tag{3}$$

(Faraday's law: another source of **E** is a changing **B**. The surface Σ is bounded by the closed curve C.)

$$\oint_C \mathbf{B} \cdot d\mathbf{l} = \epsilon\mu \frac{d}{dt} \iint_\Sigma \mathbf{E} \cdot d\mathbf{S} + \mu I \tag{4}$$

(Ampère's law: the sources of **B** are both the changing **E** on the surface Σ bounded by the closed curve C and the net current through C.)

In these equations, $d\mathbf{S}$ is an infinitesimal surface area and $d\mathbf{l}$ is an infinitesimal length. $\oint$ means that the integral goes around a closed curve. $\oiint$ means that the integral goes over a closed surface. These equations result in a *wave equation*:

$$\frac{\partial^2 \mathbf{E}}{\partial x^2} = \frac{1}{c^2} \frac{\partial^2 \mathbf{E}}{\partial t^2}, \qquad \frac{\partial^2 \mathbf{B}}{\partial x^2} = \frac{1}{c^2} \frac{\partial^2 \mathbf{B}}{\partial t^2}.$$

Equations (3) and (4) show how this occurs: a changing **E** creates some new **B**, and a changing **B** creates some new **E**. (We are thinking of empty space, where $I = 0$. The situation is more complicated when $I \neq 0$.) Thus the fact of changing fields leads to new (changing) fields, and the wave propagates itself:

$$c^2 = \frac{1}{\mu\epsilon}.$$

(Hence the index of refraction is $\sqrt{\epsilon/\epsilon_0}$ when $\mu = \mu_0$.)

We will be concerned with regions of space where $Q = I = 0$: Consider that there exists some field **E**. We will *define* the y direction as the direction of **E**. That is, $\mathbf{E} \equiv \mathbf{j}E_y$. Further, we *define* the y, z plane as identical to the **E**, **B** plane; that is, there may be B_y and B_z, but *not* B_x.

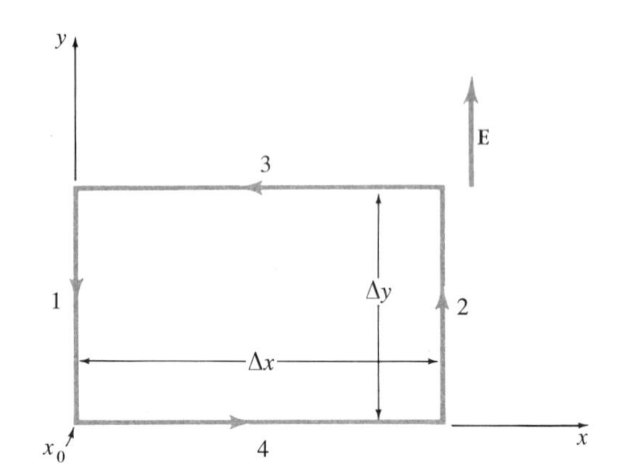

Figure D-1

Now we will apply Eq. (3) to the small rectangular curve of Figure D-1, in the x, y plane: $\mathbf{E} \cdot d\boldsymbol{l} = 0$ on sides 3 and 4. On side 1,
$E = E(x_0), \qquad dl = -\Delta y.$

On side 2,

$E = E(x_0 + \Delta x), \qquad dl = +\Delta y.$

So the left side of Eq. (3) is

$$\begin{aligned}\oint \mathbf{E} \cdot d\boldsymbol{l} &= -E(x_0) \cdot \Delta y + E(x_0 + \Delta x) \cdot \Delta y \\ &= +\Delta y \, \Delta x \left[\frac{E(x_0 + \Delta x) - E(x_0)}{\Delta x}\right] \\ &= +\Delta y \, \Delta x \frac{\partial E}{\partial x},\end{aligned}$$

suppressing the limit for the moment.

The right side of Eq. (3) is

$$-\frac{\partial}{\partial t} \int \mathbf{B} \cdot d\mathbf{S} = -\frac{\partial}{\partial t} (B \cdot \Delta x \, \Delta y).$$

Combining, and canceling $\Delta x \cdot \Delta y$, we get

$$-\frac{\partial E}{\partial x} = \frac{\partial B}{\partial t}. \tag{5}$$

Some comments: We have used the fact that the integral is the limit of a sum of small bits, and when the path itself is small, we can just use the sum. Also, the d/dt has become $\partial/\partial t$, since in our case, x, y, and z are all kept constant.

Now $E = E_y$, the only one that exists. B is the component of $\mathbf{B}$, which is perpendicular to the x, y plane; that is, $B = B_z$. Therefore

$$\frac{\partial E_y}{\partial x} = -\frac{\partial B_z}{\partial t}. \tag{5'}$$

Now repeat the process for Eq. (4), using a path in the x, z plane (so that a possible B_y is not involved). We get

$$\frac{\partial B_z}{\partial x} = -\epsilon\mu\frac{\partial E_y}{\partial t}. \tag{6}$$

Differentiate Eq. (5′) by x,

$$\frac{\partial^2 E_y}{\partial x^2} = -\frac{\partial^2 B_z}{\partial x\,\partial t},$$

and Eq. (6) by t,

$$\frac{\partial^2 E_y}{\partial t^2}\cdot\epsilon\mu = -\frac{\partial^2 B_z}{\partial x\,\partial t}.$$

Combining,

$$\frac{\partial^2 E_y}{\partial x^2} = \epsilon\mu\frac{\partial^2 E_y}{\partial t^2}. \tag{7}$$

Similarly,

$$\frac{\partial^2 B_z}{\partial x^2} = \epsilon\mu\frac{\partial^2 B_z}{\partial t^2} \tag{8}$$

Notes

1. If B_y exists, it is not governed by a wave equation, and therefore is not of the form $f(x \pm ct)$.
2. Solutions to Eqs. (7) and (8) are of the form $f(x \pm ct)$, *if* $c^2 = 1/\epsilon\mu$.

3. We defined the y direction as along **E**, and the z direction as along that component of **B** which is perpendicular to **E**. Then x, the propagation direction, is unique: $\hat{\mathbf{i}} = \hat{\mathbf{j}} \times \hat{\mathbf{k}}$. The wave $f(x - ct)$ moves toward positive x, as specified this way. Note that this way of defining direction *requires* E_y and B_z to be positive.
4. Solutions to Eqs. (7) and (8) are of the form $f(x \pm ct)$. For instance,

$$E_y(x, t) = E_0 \sin 2\pi\left(\nu t - \frac{x}{\lambda}\right). \tag{9}$$

For generality, we can write, then,

$$B_z(x, t) = B_0 \sin\left[2\pi\left(\nu t \pm \frac{x}{\lambda}\right) + \phi\right]. \tag{10}$$

To find which sign should be used, apply Eqs. (5′) and (6):

$$-\left(\frac{2\pi}{\lambda}\right)E_0 \cos 2\pi\left(\nu t - \frac{x}{\lambda}\right) = 2\pi\nu B_0 \cos\left[2\pi\left(\nu t \pm \frac{x}{\lambda}\right) + \phi\right] \tag{11}$$

and

$$-2\lambda\nu\epsilon\mu E_0 \cos 2\pi\left(\nu t - \frac{x}{\lambda}\right) = \pm\left(\frac{2\pi}{\lambda}\right)B_0 \cos\left[2\pi\left(\nu t \pm \frac{x}{\lambda}\right) + \phi\right].$$

Dividing, we get

$$\frac{(2\pi/\lambda)}{\epsilon\mu 2\pi\nu} = \pm\,\nu\lambda;$$

since $\nu\lambda = c$ and $c^2 = 1/\epsilon\mu$, we see that only the solution with $[2\pi(\nu t - x/\lambda) + \phi]$ makes sense.
5. Consider Eq. (11), which is true for all values of t and x at the points where $\nu t - x/\lambda = \frac{1}{4}$:

$$-\left(\frac{2\pi}{\lambda}\right)E_0 \cdot 0 = -2\pi\nu B_0 \cos\left(\frac{\pi}{2} + \phi\right) = -2\pi\nu B_0 \sin(\phi).$$

So $\sin\phi = 0$, which means that $\phi = 0$ or π. Next consider Eq. (11) at the points where $\nu t - x/\lambda = 0$:

$$-\left(\frac{2\pi}{\lambda}\right)E_0 \cdot 1 = -2\pi\nu B_0 \cos\phi.$$

Since λ, E_0, ν, and B_0 are all positive, $\phi = 0$, not π. Thus, the solution compatible with Eq. (9) is

$$B_z(x, t) = B_0 \sin 2\pi\left(\nu t - \frac{x}{\lambda}\right), \qquad B_0 = \frac{E_0}{c}. \tag{12}$$

Now what about the (possible) solution

$$E_y(x, t) = E_0 \sin 2\pi\left(\nu t + \frac{x}{\lambda}\right)? \tag{13}$$

The same sort of arguments as given above require that B have a solution of the form

$$\begin{aligned} B_z(x, t) &= \left(\frac{E_0}{c}\right) \sin\left[2\pi\left(\nu t + \frac{x}{\lambda}\right) + \pi\right] \\ &= -\left(\frac{E_0}{c}\right) \sin 2\pi\left(\nu t + \frac{x}{\lambda}\right). \end{aligned} \tag{14}$$

Figure D-2

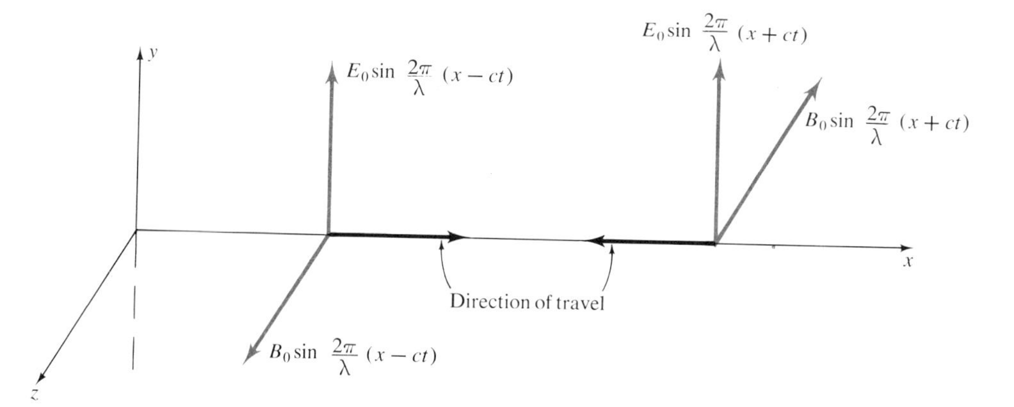

From this we get a picture like Figure D-2. Notice that in both cases the direction of travel is the direction of $\mathbf{E} \times \mathbf{B}$.

Appendix E: Resonance

When an electromagnetic wave encounters a charge (as it does in matter of any sort), the charge is accelerated by the electric field. We know that accelerated charges radiate, and regard this as the source of the "scattered" electromagnetic wave. We encounter this situation in the discussion of light pressure, and use it again to discuss the index of refraction. The present discussion will explain the phase shift of the reradiated wave, which governs both these effects and the accompanying absorption of energy.

Let us write down all the forces acting on the charge, and set their sum equal to ma (at $x = 0$):

$$qE \cdot \sin 2\pi\nu t - \beta v_y - ky = m\frac{d^2y}{dt^2}.$$

If we write $v_y = dy/dt$, then the equation governing y is the standard differential equation for a forced, simple-harmonic oscillator:

$$\frac{d^2y}{dt^2} + \frac{\beta}{m}\frac{dy}{dt} + \frac{k}{m}y = \frac{qE_0}{m}\sin 2\pi\nu t.$$

The solution to this is $y = A\sin(2\pi\nu + \phi)$, which may be expanded as

$$y = A\cos\phi\sin 2\pi\nu t + A\sin\phi\cos 2\pi\nu t.$$

We can then define the "susceptibilities", χ' and χ'' so that

$$y = (\chi' \sin 2\nu t + \chi'' \cos 2\nu t)E_0 .$$

We can solve for A and ϕ, or for χ' and χ''.

Two algebraic equations result from plugging the solution into the differential equation, since terms proportional to $\sin 2\pi\nu t$ are independent of those proportional to $\cos 2\pi\nu t$. (To see this, set $t = 0$ and find coefficients of $\cos 2\pi\nu t$, since the equation is true at all time, and the coefficients are constants. Then do the same for $t = \tau/4$). We find

$$\chi' = \frac{(-q_0/m)(\omega^2 - (k/m))}{(\omega^2 - (k/m))^2 + (\beta\omega/m)^2} \simeq \frac{-q_0}{2m\omega} \frac{\Delta\omega}{\Delta\omega^2 + (\beta/2m)^2}$$

$$\chi'' = \frac{(-q/m)(\omega\beta/m)}{(\omega^2 - (k/m))^2 + (\beta\omega/m)^2} \simeq \frac{-q\beta}{2m^2\omega} \frac{1}{\Delta\omega^2 + (\beta/2m)^2}$$

where $\omega = 2\pi\nu$, $\Delta\omega = 2\pi\ \Delta\nu = \omega - \omega_{res}$. $\omega_{res} = \sqrt{k/m}$. The approximation holds near the resonant frequency, that is, when $\omega \simeq \omega_{res}$. In the other form, we note that $A^2 = (\chi'^2 + \chi''^2)E_0{}^2$ and $\tan\phi = \chi''/\chi'$. These results are generally summed up in Figure E-1

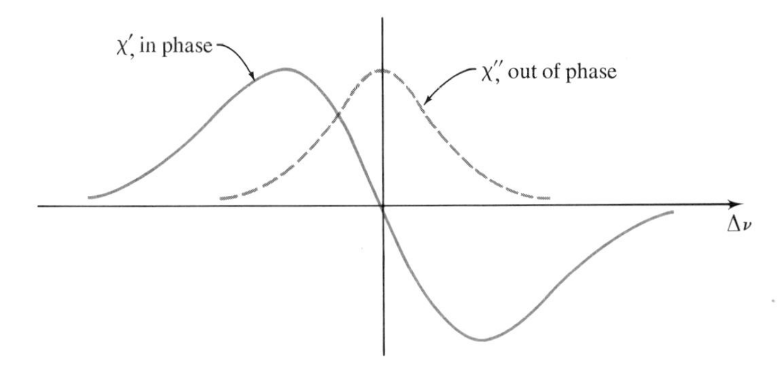

Figure E-1

by graphs of the in-phase and out-of-phase components of y, in the region near $\Delta\nu = 0$.

Since χ'' is the coefficient of the component of y which is out of phase with the driving force, it must be the coefficient of the in-phase component of v_y. Therefore the work done by this force (that is, the energy absorbed by the bound charge) is proportional to χ'':

$$W = \int_0^T \mathbf{F} \cdot \mathbf{v}\, dt = \frac{qE_0{}^2}{m} \chi'' \cdot 2\pi\nu \int_0^T (\sin 2\pi\nu t)^2\, dt.$$

(Remember that

$$\int_0^T \sin 2\pi\nu t \cos 2\pi\nu t \, dt \to 0 \qquad \text{and} \qquad \int_0^T \sin^2 2\pi\nu t \to \tfrac{1}{2}$$

as T becomes large compared to τ). Thus the energy absorbed is proportional to χ''. This is why χ'' is referred to as the *absorption*.

The other component, known as the *dispersion*, refers to the re-radiated energy. For a single charge the re-radiated wave follows the acceleration of the charge, which has the same phase as y. When there are many charges, the re-radiated wave has the phase of v, $\pi/2$ different from ϕ, which is the phase of y:

$$\tan \phi = \frac{\chi''}{\chi'} = \frac{4\pi\,\Delta\nu \cdot m}{\beta}.$$

Figure E-2

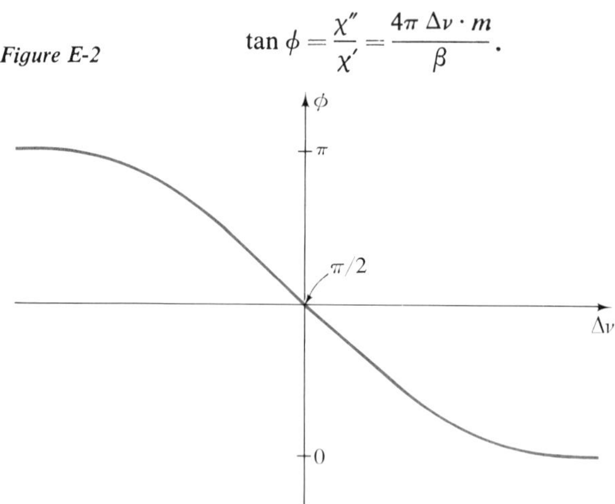

Note first that if there are no frictional forces, there is no absorption of energy and no phase shift. Second, if $\Delta\nu$ is large (the incident radiation has a frequency far above or below the resonant frequency),

Figure E-3

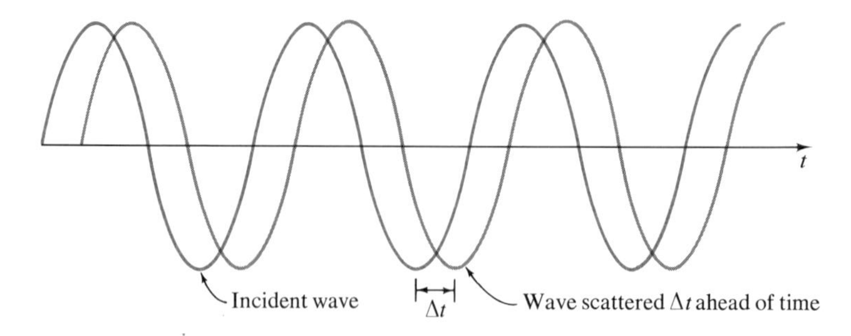

the energy will be re-radiated with phase change $\pm\pi/2$ (for many charges) and without absorption. This amounts to no detectable change.

If we think of the phase shifts as representing time lags (or leads!) in the re-emission process, we can draw the waves as shown in Figure E-3. The wave scattered with some nonzero phase shift appears to have gained or lost some time. This is the origin of the frequency dependence of the wave speed, which we wish to study.

Figure E-4

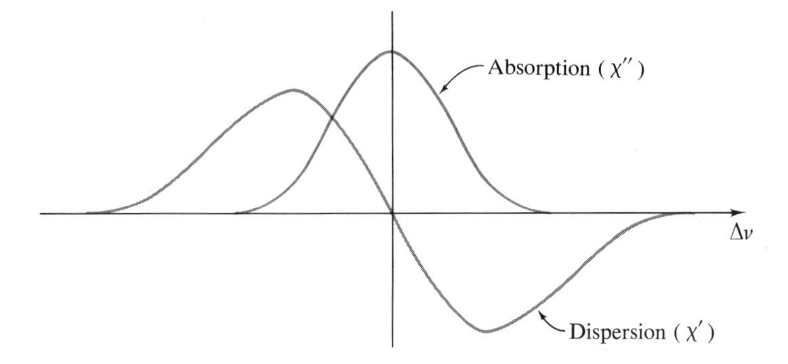

Appendix F: Index of refraction — macroscopic approach

The actual form of the electromagetic wave scattered by charges in matter is not so simple as that from a single charge. We have seen that phase shifts are critical here, and it turns out that an additional shift of $\pi/2$ is introduced when we consider the whole infinite sheet of charges at any one depth in the material. A way to avoid this is to use a macroscopic approach, that is, one with the averaging built in. Such an approach is implicit in the theory of electricity and magnetism.

Our charges, q, are all displaced from their equilibrium positions by a distance y. Since, at equilibrium, strict charge neutrality exists, the charges must have left equivalent charges of opposite sign behind (ion cores in most matter). So dipoles are set up, of magnitude $\xi q y$, where ξ is a constant depending on the material. If there are N charges per unit volume, the dipole moment per unit volume is $\xi N q \mathbf{y} = \mathbf{P}$, the polarization. We then write $\mathbf{P} = \chi_e \mathbf{E}$, where χ_e is called the electric susceptibility. $\mathbf{D} = \epsilon \mathbf{E} = \epsilon_0 \mathbf{E} + \mathbf{P}$, so $\epsilon = (1 + \chi_e)\epsilon_0$, and it is for ϵ which we need to find the index:

$$n^2 = \frac{c^2}{v_w^{\,2}} = \frac{\epsilon\mu}{\epsilon_0 \mu}.$$

Thus, we should find ϵ in terms of y, which we already know:

$$\epsilon' = \epsilon_0(1 + \xi Nq\chi'), \qquad \epsilon'' = \xi Nq\chi''\epsilon_0,$$

taking χ' and χ'' from the discussion of resonance. The reason we have an ϵ'' term is the same as the reason for χ'', and the result is again absorption. We write the full dispersion equation as

$$n^2 = 1 + \frac{(\xi Nq^2/2m\omega\epsilon_0)(\omega_{\text{res}} - \omega)}{\Delta\omega^2 + (\beta/2m)^2}.$$

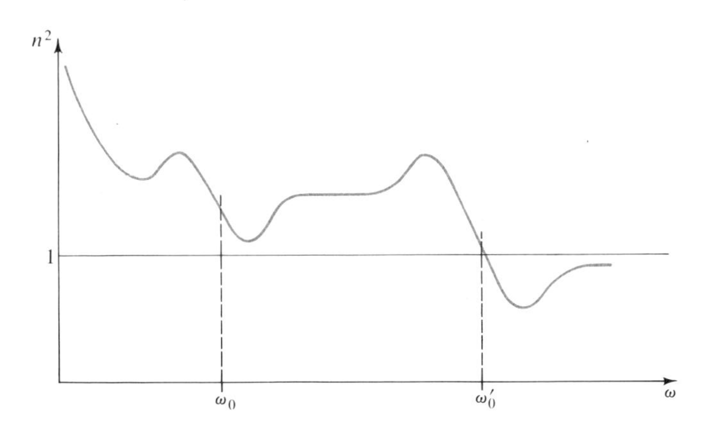

Figure F-1

Figure F-1 shows n^2 with two resonances: In a metal, $\omega_{\text{res}} \simeq 0$, since charges are nearly free, and we can then get

$$n^2_{\text{metal}} = 1 - \frac{\xi Nq^2}{2m[\omega^2 + (\beta/2m)^2]\epsilon_0}.$$

When $n^2 < 1$, the index is imaginary, and the metal reflects perfectly. As predicted here, metals do become transparent at high enough frequencies.

Appendix G: Fresnel diffraction

So far we have always made our observations of interference effects in the Fraunhofer limit; that is, where all waves are far enough from their sources to be plane. If we relax this restriction, we will be in the realm of *Fresnel diffraction* (diffraction being the general term to cover diffraction and interference effects). We consider very briefly what modifications should be introduced. There are principally two:

1. A geometric one due to the fact that we can no longer regard two rays from the source points as parallel.
2. The fact that the intensities in the interfering beams may differ, due to the variation of distance from the source, means that we must take into account the exact form of the wave because the intensity in a spherical wave falls off as $1/r^2$ and that in a cylindrical wave (from a slit source) falls off as $1/r$. (Plane sources, like those used in amplitude division, have no variation of intensity with distance—if the plane is truly infinite.)

G.1 *Two slits*

$$E_1 = E_0 \frac{1}{\sqrt{r_1}} \sin\left(\frac{2\pi r_1}{\lambda} - 2\pi\nu t + \phi\right),$$

$$E_2 = E_0 \frac{1}{\sqrt{r_2}} \sin\left(\frac{2\pi r_2}{\lambda} - 2\pi\nu t + \phi\right).$$

As usual, we can easily determine the value of E_T at any specified point. To find the points of maximum intensity, however, is more difficult, even in this simplest case. Consider the phasor in Figure

Figure G-1

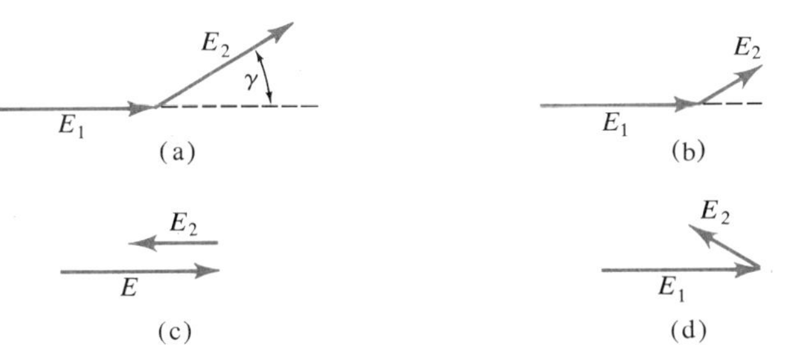

G-1(a): when $r_1 \simeq r_2$, $I_T = I_0 \cos^2 \gamma/2$, $\gamma = 2\pi(r_1 - r_2)/\lambda$. But if $r_2 \gg r_1$, then we have Figure G-1(b). But E_T in Figure G-1(c) may be larger than E_T in Figure G-1(d). The pattern in the plane perpendicular to the slits is shown in Figure G-2(a). The lines of $\Delta L = m\lambda$ are hyperbolas.

Figure G-2

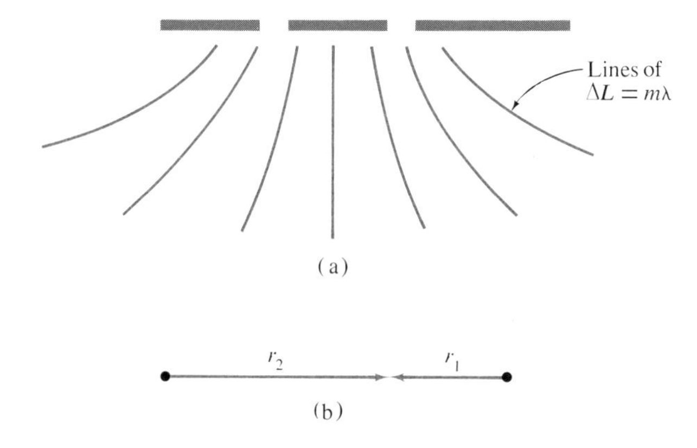

The extreme effect of the $1/\sqrt{r}$ dependence is between sources, as shown in Figure G-2(b). For instance, if $r_2 = 9\lambda$ and $r_1 = 1\lambda$, then $E_1 = 3E_2$ and $E_T = 4E_2$ (here, $\gamma = 0$). On the other hand, at $r_1 = \lambda/2$, $r_2 = 9.5\lambda$,

$$E_1' = -19E_2' = 1.4E_1.$$

Here, $\gamma = \pi$, so

$$E_T = (1.4)3E_2 - \left(\frac{\sqrt{18/19}}{\sqrt{19}}\right)E_2 = 3\sqrt{2}E_2 - \frac{3\sqrt{2}E_2}{19} = \left(\frac{(3\sqrt{2})18}{19}\right)E_2.$$

Thus the point with $\gamma = \pi$ is brighter than that with $\gamma = 0$! This is very complicated and becomes more so when we do the general problem. Here the phasors add up to parts of a complicated curve known as a *Cornu spiral*. The procedure is involved, and merits use of computers in actual problems.

G.2 *Single slit*

What problem *can* we solve then? As usual, simple geometries lend themselves to the location of salient features of the interference pattern — the centers of the bright and dark fringes. Consider the problem presented by Figure G-3. What fringe do we see along the normal line through the center of a single slit?

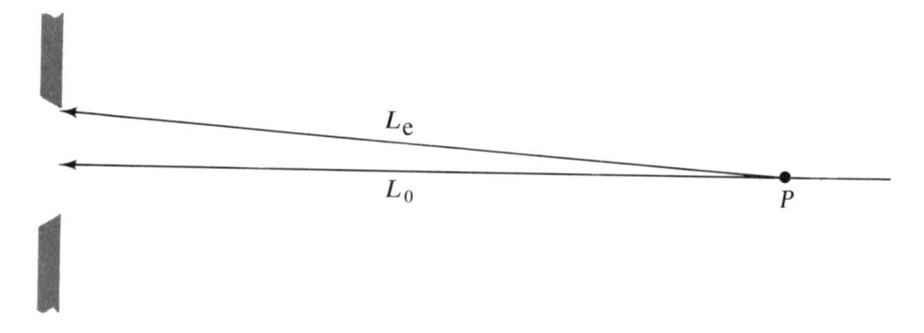

Figure G-3

If $L_e - L_0 \ll \lambda$, then P is bright; that is, all light reaches it approximately in phase (Fraunhofer limit). If $L_e - L_0$ is not small compared with λ, then we may divide the slit into "zones":

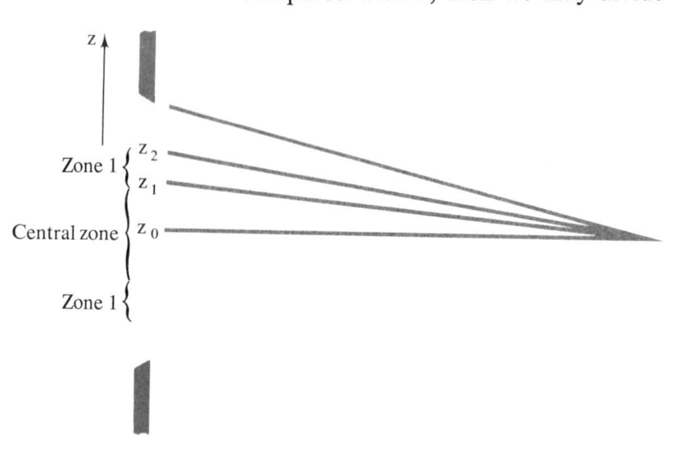

Figure G-4

$$L_1 = L_0 + \frac{\lambda}{4}, \qquad L_2 = L_1 + \frac{\lambda}{2}, \qquad L_3 = L_2 + \frac{\lambda}{2}, \qquad \text{etc.}$$

The central zone contributes waves relatively in phase, say $\phi = 0$. The *next* zone contributes waves all out of phase with respect to zone 1, and so cancels the amplitude from zone 1. The third adds to the first, the fourth to the second, and so forth. The number of zones present depends on L_0:

When far away, the whole slit belongs to the central zone, and P is a bright spot.

Somewhere closer, there will be just two zones, and P is dark.

Closer yet, there are three zones, and P is bright; and so forth.

Suppose we mask off every other zone; then we get a bright spot (no matter whether we mask odd or even zones. Compare Babinet's principle.) This is a Fresnel *zone plate.*

The zone plate functions somewhat as a lens, since plane waves falling on it create a bright spot at a single point. One can think of this as a primitive hologram.

One historically important confirmation of Fresnel's theory is the appearance of a bright spot in the center of the shadow of a disk (or sphere; ball bearings are often used). Because of the $1/r$ variation, each zone contributes a slightly smaller wave than the previous zone. The disk masks the first N zones, and the $(N+1)$ contributes an amplitude E_{N+1} on the axis. The contribution of the $(N+2)$th does not quite cancel this. The $(N+3)$th adds to the $(N+1)$th, and the $(N+4)$th again fails to cancel it completely, and so forth. Figure G-5 shows the "shadow" of a ball bearing that is illuminated with (parallel) laser light.

Figure G-5

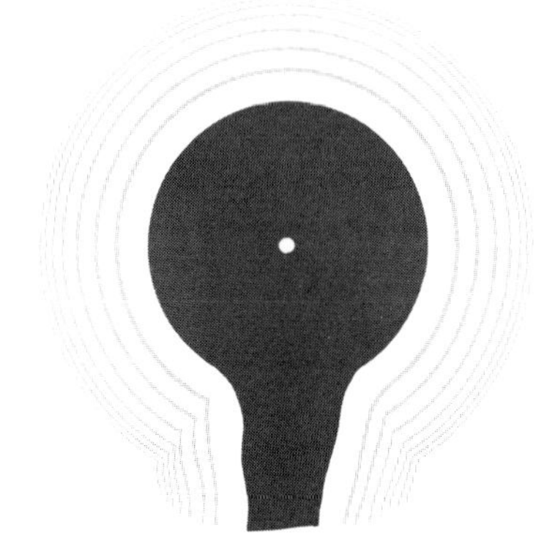

Appendix H: Fourier transforms

We have seen that if we add two waves, with frequencies ν_0 and $\nu_0 + \delta\nu$, we get a pattern of beats which travel at the group velocity $\delta\nu/\delta(1/\lambda)$. If we add a third wave, of frequency $\nu_0 - \delta\nu$, the pattern is that shown in Figure H-1. The modulation envelope here is the same

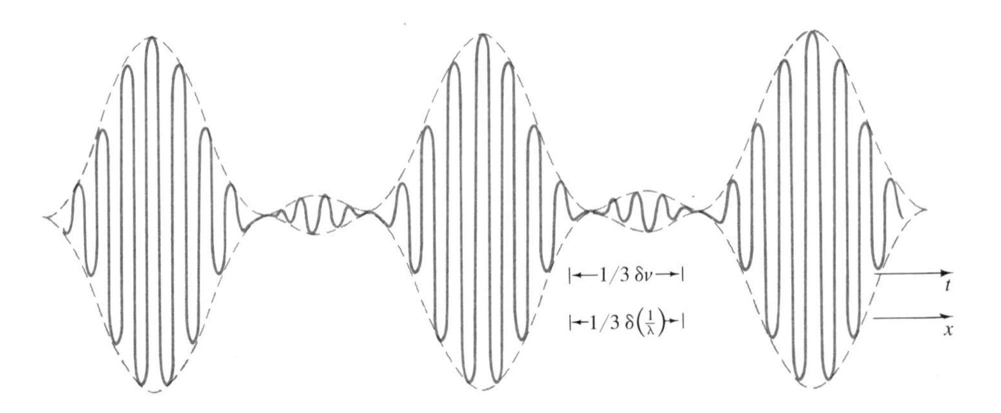

Figure H-1

as that of the curve $E_T(\gamma)$ for a three-slit interference problem. In fact, if we have $2N + 1$ waves whose frequencies differ by multiples of $\delta\nu$ we get the pulses shown in Figure H-2.

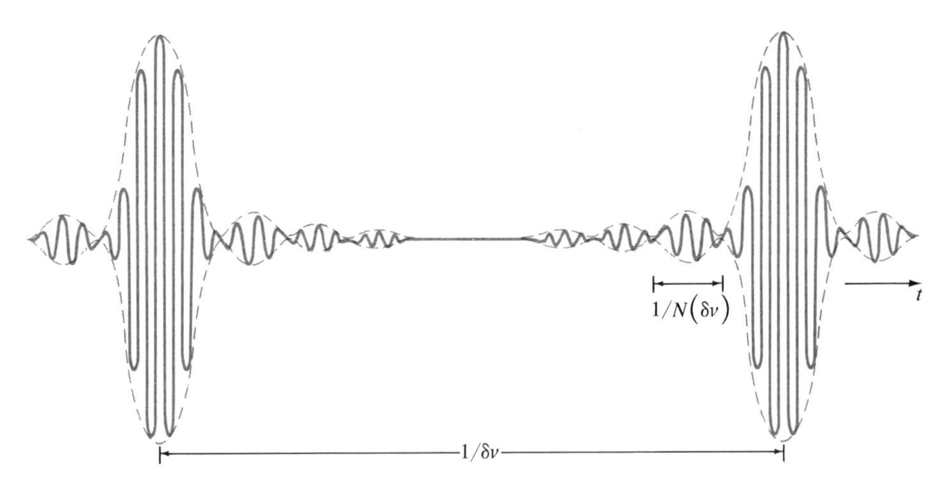

Figure H-2

If we plot a curve of intensity versus frequency, we get that of Figure H-3, where $2N + 1 = 9$. This is called a *spectral density curve,*

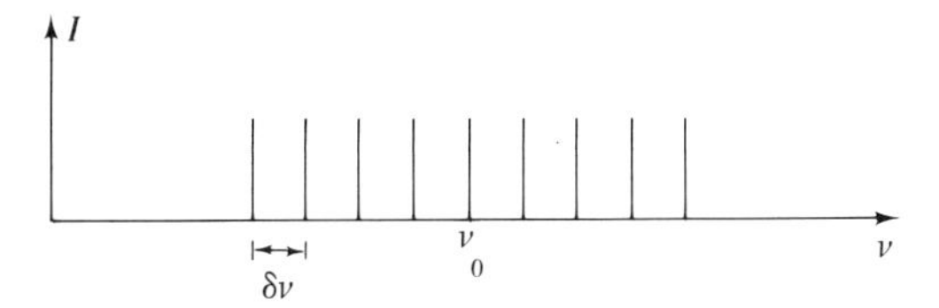

Figure H-3

or simply a spectrum, and is just a different way of displaying the information contained in Figure H-2. The term "spectral density" refers to the number of component waves per unit range of frequency. Here we could say that there is an average spectral density of $n(\nu) = 1/\delta\nu$ between $\nu_0 - N\,\delta\nu$ and $\nu_0 + N\,\delta\nu$, and $n(\nu) = 0$ elsewhere. But for such discrete waves the average does not tell the whole story.

We can add up the $2N + 1$ waves by means of the series

$$\sum_{k=-N}^{N} \sin(\theta + k\gamma) = \sin\theta \, \frac{\sin(2N+1)\gamma/2}{\sin\gamma/2},$$

which we have proved geometrically for the phasors from $2N + 1$ slits (in Chapter 10). We could also prove it as follows:

$$\sin\frac{\gamma}{2} \sum \sin(\theta + k\gamma) = \sin\theta \sum \sin\frac{\gamma}{2}\cos k\gamma + \cos\theta \sum \sin\frac{\gamma}{2}\sin k\gamma.$$

If we expand both series and use the trigonometric identities for sin A cos B and sin A sin B, we find that all terms cancel in pairs, except sin θ sin$(2N + 1)\gamma/2$.

This means that the $2N + 1$ waves of Figure H-3 add up to the pulses of Figure H-2:

$$\sum_{k=-N}^{N} \sin 2\pi\left[\left(\frac{x}{\lambda_0} - \nu_0 t\right) + k\left(x\,\delta\left(\frac{1}{\lambda}\right) - \delta\nu \cdot t\right)\right]$$

$$= \sin 2\pi\left(\frac{x}{\lambda_0} - \nu_0 t\right) \frac{\sin[(2N+1)(\delta(1/\lambda)x - \delta\nu \cdot t)\pi]}{\sin[\pi(\delta(1/\lambda)x - \delta\nu \cdot t)]}$$

This way of making up complicated curves from a set of pure sine waves is called *Fourier synthesis*, and the reverse process is called *Fourier analysis*. The commonest instances of this involve the analysis of complicated periodic functions by means of the Fourier series:

$$F_{\text{comp}}(x) = \sum_n \left(A_n \sin\frac{nx}{\Lambda} + B_n \cos\frac{nx}{\Lambda}\right),$$

where Λ is the period of the function. The coefficients are obtained by multiplying both sides of the equation by $\sin(mx/\Lambda)$ or $\cos(mx/\Lambda)$, and using the fact that

$$\int_0^{\Lambda} \sin\frac{mx}{\Lambda} \cos\frac{mx}{\Lambda}\,dx = 0,$$

and

$$\int_0^{\Lambda} \binom{\sin}{\cos} \frac{mx}{\Lambda} \binom{\sin}{\cos} \frac{nx}{\Lambda} = \begin{cases} 0 & \text{if} \quad m \neq n\text{, or} \\ 1 & \text{if} \quad m = n \end{cases}$$

For example,

$$A_m = \int_0^{\Lambda} F_{\text{comp}}(x) \sin\frac{mx}{\Lambda}\,dx.$$

Series of the Fourier type can be written for any functions that exhibit this "orthogonal" property.

Figure H-4

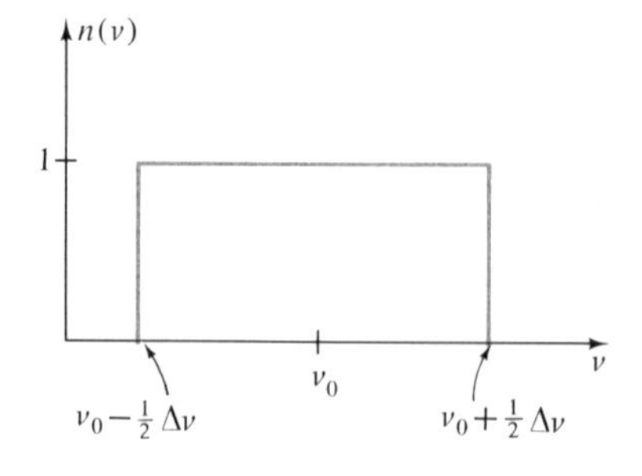

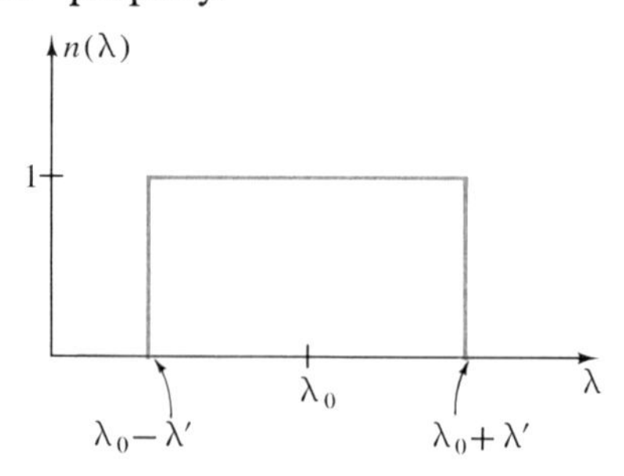

Let us add more and more component waves, all within the range $\nu_0 - \frac{1}{2}\Delta\nu$ to $\nu_0 + \frac{1}{2}\Delta\nu$. This is equivalent to letting $N \to \infty$ and $\delta\nu \to 0$, so that the curve of Figure H-3 becomes that of Figure H-4. We must also change our sum to an integral:

$$F(x, t) = \int_{\nu_0-(1/2)\Delta\nu}^{\nu_0+(1/2)\Delta\nu} \sin 2\pi\left(\frac{x}{\lambda} - \nu t\right) d\nu$$

$$= \Delta\nu \sin 2\pi\left(\frac{x}{\lambda_0} - \nu_0 t\right) \frac{\sin \pi[x\,\Delta(1/\lambda) - \Delta\nu t]}{\pi[x\,\Delta(1/\lambda) - \Delta\nu t]},$$

which we can confirm by direct integration or by reference to the single-slit problem of Chapter 12. The pulse, of course, looks like

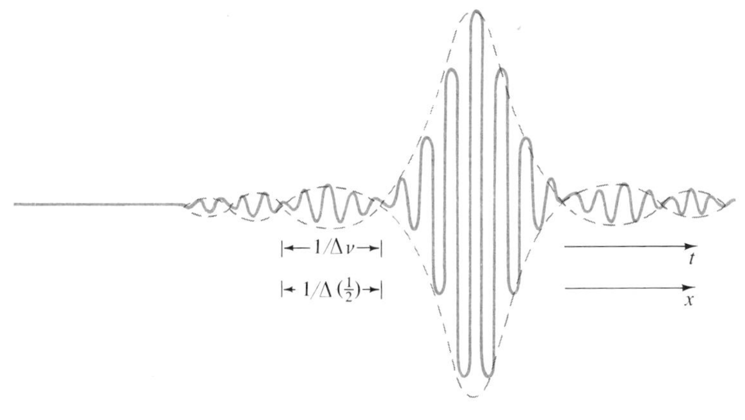

Figure H-5

that shown in Figure H-5: It is not periodic; that is, only one pulse exists.

Now, we could also have written our integral as

$$\int_0^\infty n(\nu) \sin 2\pi\nu t\, d\nu = F(t).$$

Written this way, $F(t)$ is called the Fourier sine transform of $n(\nu)$. In Chapter 12, $E_T(\gamma)$ is the Fourier sine transform of $E(\gamma)$, the function which describes the slit or other diffracting object. Again we use the orthogonality of $\sin 2\pi\nu t$ to show that

$$n(\nu) = \frac{1}{\pi}\int_{-\infty}^{\infty} F(t) \sin 2\pi\nu t\, dt.$$

That is, the transform of a transform is the original function. We say that $n(\nu)$ is the inverse transform of $F(t)$. The parameters x and $1/\lambda$ also form a pair like t and ν.

We can use this property of inversion to find out how "pure" a sine wave can be. No wave train is infinitely long, and this fact keeps it from being strictly monochromatic, a situation we have found experimentally with the interferometer. Suppose we have a sine wave extending over a distance L_C as shown in Figure H-6. We can describe

Figure H-6

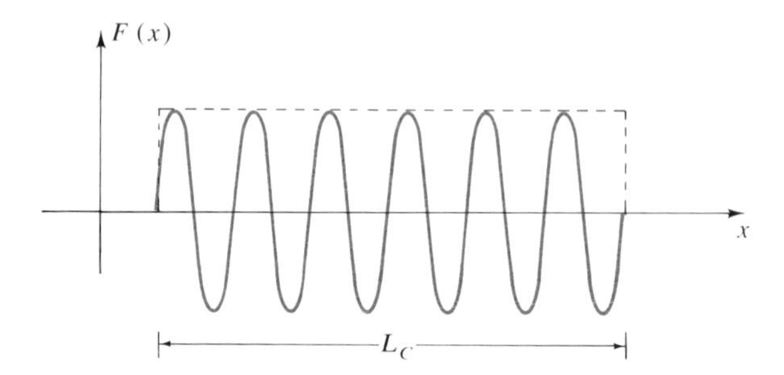

this as the product of two functions: $A(x) \cos 2\pi(x/\lambda_0)$, where $A(x)$ is the same rectangular function as $n(\nu)$ in Figure H-4. Without even doing the mathematics, we know that the spectral components of this wave are those shown in Figure H-7. The larger L_C, the narrower the $n(\lambda)$ curve, and the more monochromatic the wave.

Figure H-7

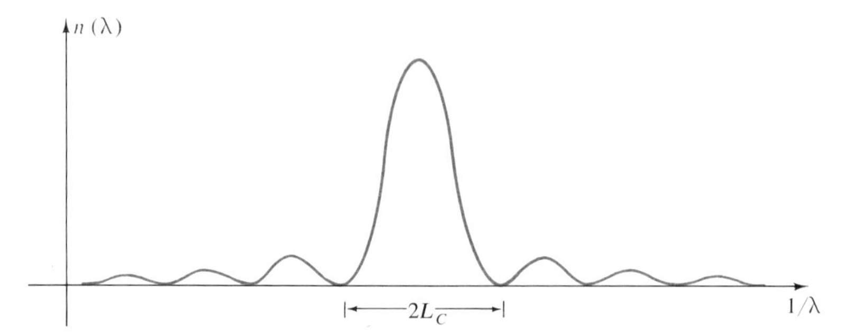

The pervasive generality of Fourier transforms in physics is suggested by the two rather different physical situations which have led us to the same mathematical device.

A cosine transform also exists, corresponding to the cosine terms in the series, as well as a general transform which is a linear combination of the two, using $e^{i\theta} = \cos\theta + i\sin\theta$ notation.

The student is referred to the Bibliography for more extensive treatments of this important subject. What we have seen here is that a dualism exists between two complementary ways of organizing the information given by a function. This dualism is exhibited by the mathematical device of Fourier analysis or transformation, and occurs physically in diffraction and the harmonic analysis of nonharmonic (nonsinusoidal) waves.

Bibliography

Jenkins, F. A., and White, H. E., *Fundamentals of Optics*, 3d ed. (1957). McGraw-Hill, New York. The standard reference on classical optics. The authors have actually carried out many of the operations discussed, and the practical details are invaluable.

Stone, J. M., *Radiation and Optics* (1963). McGraw-Hill, New York. A modern advanced treatment. Particularly good on scattering theory and Fourier transforms.

Strong, J., *Concepts of Classical Optics* (1958). W. H. Freeman, San Francisco, Practical details for experimentation.

Goldwasser, E. L., *Optics, Waves, Atoms, and Nuclei* (1965). W. A. Benjamin, New York. Good elementary treatment, but rather brief.

Feynman, R. P., Leighton, R. B., and Sands, M., *The Feynman Lectures on Physics*, Vol. I (1964). Addison-Wesley, Reading, Mass. Brief but insightful. Useful, particularly on index of refraction and the relationship of optical interference to quantum mechanics.

Fowles, G. R., *Introduction to Modern Optics* (1968). Holt, Rinehart and Winston, New York. An advanced text, thoroughly up to date.

Francon, M., *Optical Interferometry* (1966). Academic Press, New York. Very advanced, but entirely complete on most of modern optics. Quite readable.

Shurcliff, W. A., and Ballard, S. S., *Polarized Light* (1964). Van Nostrand (Momentum), Princeton, N.J. Specialized, but usable.

Wood, R. W., *Physical Optics* (1934). Macmillan, New York. The ultimate authority on many early experiments.
Born, M., and Wolf, E., *Principles of Optics* (1964). Macmillan, New York. Ponderous, but complete.
Tolansky, S., *High Resolution Spectroscopy* (1947). Pitman, New York. On interferometry.
Also:
Rossi, B., *Optics* (1957). Addison-Wesley, Reading, Mass.
Towne, D. H., *Wave Phenomena* (1967). Addison-Wesley, Reading, Mass.
Ditchburn, R. W., *Light* (1963). Wiley, New York.
Young, H. D., *Optics and Modern Physics* (1968). McGraw-Hill, New York.

Articles of interest to advanced students: These are all taken from *American Journal of Physics*, a journal publishing mostly explanatory and review articles on topics of current interest. Where the title does not make it obvious, the relevant chapters of this text are given in parentheses.

Weidner, R. T., *Am. J. Phys.* **35**, 443 (1967); *On weighing photons.*
deWitte, A. J., *Am. J. Phys.* **35**, 301 (1967); *Interference in scattered light.*
Givens, M. P., *Am. J. Phys.* **35**, 1056 (1967); *Introduction to holography.*
Jeppesen, M. A., *Am. J. Phys.* **35**, 435 (1967); *Measurement of dispersion of gases with a Michelson interferometer.*
Albergotti, J. C., *Am. J. Phys.* **35**, 1092 (1967); *Instant holograms.*
Apparatus Notes Section, *Am. J. Phys.* **35**, xxii (1967); *White-light interference fringes* and *Inexpensive Fabry-Perot interferometer.*
Webb, R. H., *Am. J. Phys.* **36**, 62 (1968); *Holography for the sophomore laboratory.*
Greenberg, J. M., and Greenberg, J. L., *Am. J. Phys.* **36**, 274 (1968); *Does a photon have a rest mass?*
King, A. L., *Am. J. Phys.* **36**, 456 (1968); *Multiple laser beams for the elementary laboratory.*
Kittel, C., *Am. J. Phys.* **36**, 610 (1968); *X-ray diffraction from helices: structure of DNA.*
Pollock, R. E., *Am. J. Phys.* **31**, 901 (1963); *Resonant detection of light pressure by a torsion pendulum in air.* See also G. B. Brown's comment, *Am. J. Phys.* **34**, 272 (1966).
Streib, J. F., *Am. J. Phys.* **34**, 1197 (1966); *Laws of reflection and refraction: photon flux.*
Tea, P. L., Jr., *Am. J. Phys.* **33**, 190 (1965); *Some thoughts on radiation pressure.*
Weltin, H., *Am. J. Phys.* **33**, 413 (1965); *Useful surplus item for the physical optics laboratory* (Chap. 6).
French, A. P., and Smith, J. H., *Am. J. Phys.* **33**, 532 (1965); *Low-cost Fabry-Perot interferometers.* See also H. W. Wilson, p. 1090, and F. M. Phelps III, p. 1088, of the same volume.
Bloor, D., *Am. J. Phys.* **32**, 936 (1964); *Coherence and correlation—two advanced experiments in optics.* See also the comment by Hindmarsh and King, *Am. J. Phys.* **33**, 968 (1965).

Halbach, K., *Am. J. Phys.* **32**, 90 (1964); *Matrix representation of Gaussian optics* (Chap. 1). There is also considerable comment on this (mostly on notation) in later volumes.

Dutton, D., Givens, M. P., and Hopkins, R. E., *Am. J. Phys.* **32**, 355 (1964); *Some demonstration experiments in optics using a gas laser.*

Young, P. A., *Am. J. Phys.* **32**, 367 (1964); *A student experiment in Fresnel diffraction.*

Marthienssen W., and Spiller, E., *Am. J. Phys.* **32**, 919 (1964); *Coherence and fluctuations in light beams.*

Solutions to selected problems

1.2 We apply the rule for refraction at each surface:

Surface 1: $n_1 \sin \theta_1 = n_2 \sin \theta_2$,

Surface 2: $n_2 \sin \theta_2 = n_3 \sin \theta_3$,

Surface 3: $n_3 \sin \theta_3 = n_1 \sin \theta_1'$.

Rather simply, we see that $\theta_1 = \theta_1'$; so the beam emerges at 45 degrees.

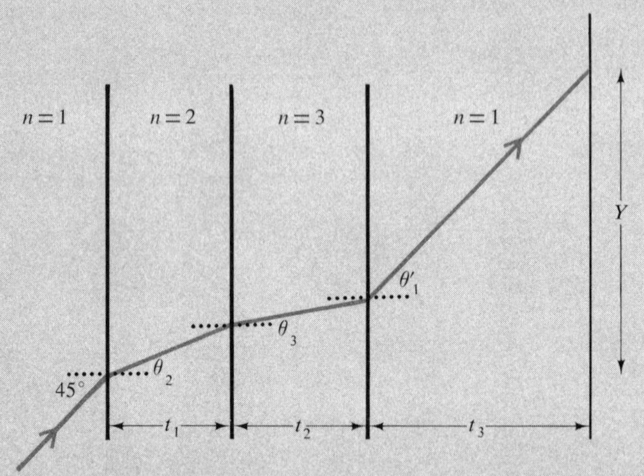

We must actually do the calculation to find the distance y:

$$y = t_1 \tan \theta_2 + t_2 \tan \theta_3 + t_3 \tan 45°.$$

Next include the reflected beams: The drawing shows the first three beams, and it is apparent that the third one is brighter than the second. The fourth will, of course, be less bright again.

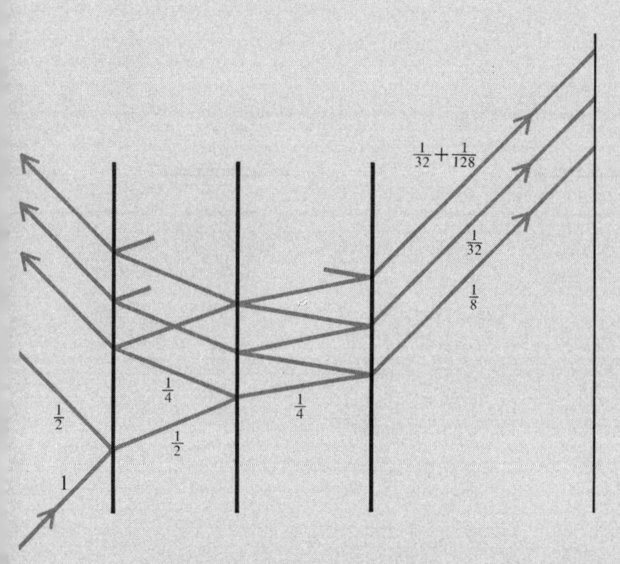

1.4 (a) The beam will be totally internally reflected, since

$\frac{4}{3}\sin 60° > 1.$

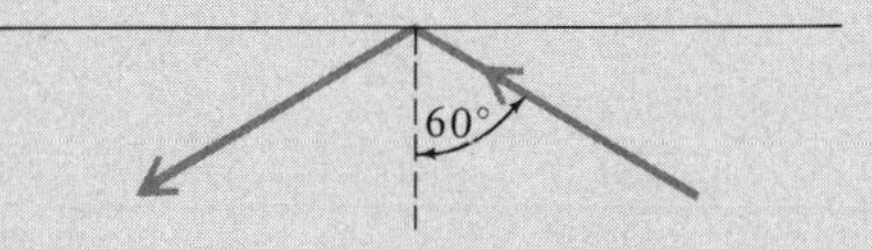

(b) Now the beam penetrates the oil layer and is in turn totally reflected from its interface with air, since

$\frac{4}{3}\sin 60° = 1.152 = 1.2\sin\theta > 1.$

(c) The same argument leads to

$$\frac{4}{3}\sin 60° = \frac{2}{\sqrt{3}} > 1.15.$$

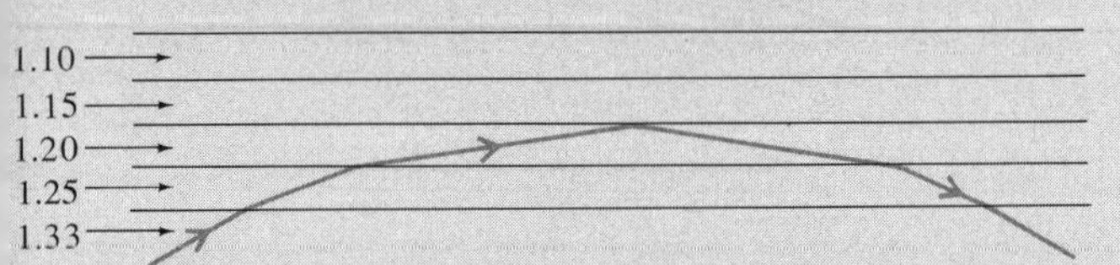

(d) The index of the air changes continuously rather than in discrete steps. Otherwise, the behavior is the same as that in part (c). Total reflection finally occurs at an index given by $1.0003\sin 89° = n$. So $n = 1.0001$.

1.6 (a) For the simple magnifier:

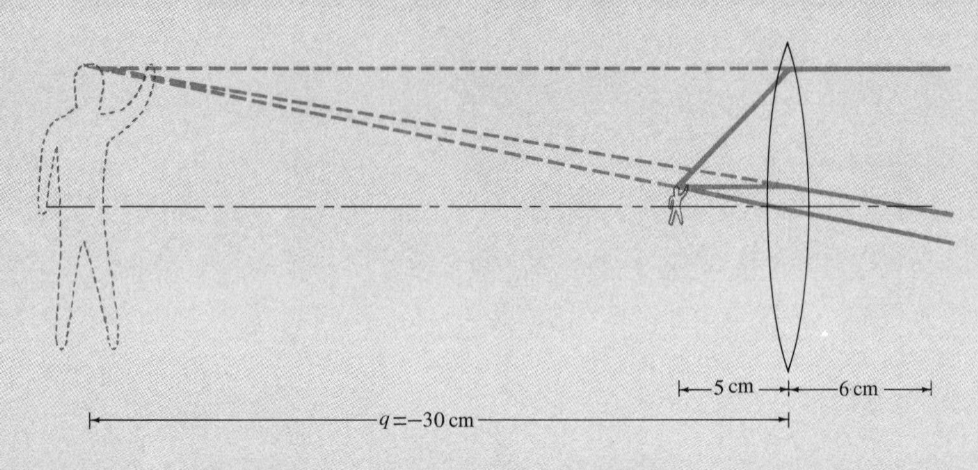

$$\frac{1}{5}+\frac{1}{q}=\frac{1}{6} \qquad q=-30 \text{ cm} \qquad \text{(virtual)},$$

$$Y_{\text{im}}=-\left(\frac{q}{p}\right)Y_{obj}=-\left[\frac{-30}{5}\right]1 \text{ mm}=+6 \text{ mm} \qquad \text{(erect)}.$$

(b) For the compound microscope:

$$q_1=\frac{f_1 p_1}{p_1-f_1}=\frac{1.1}{0.1}=+11 \text{ cm} \qquad \text{(real)},$$

$$Y'=-\frac{11}{1.1}(0.01)\text{mm}=-0.1 \text{ mm} \qquad \text{(inverted)},$$

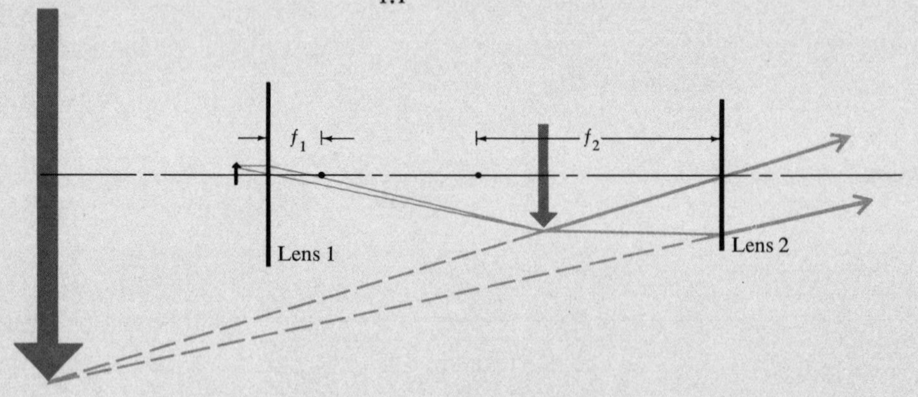

$$p_2 = L - q_1 = 18 - 11 = +7 \text{ cm} \qquad \text{(real)},$$

$$q_2 = \frac{f_2 p_2}{p_2 - f_2} = \frac{10(7 \text{ cm})}{7 - 10} = \frac{-70}{3} \text{ cm} = -23.3 \text{ cm} \qquad \text{(virtual)},$$

$$Y_2 = -\left(\frac{-23.3}{7}\right)(-0.1 \text{ mm}) = -\frac{1}{3} \text{ mm} \qquad \text{(inverted)}.$$

(c) For the astronomical microscope:

$$q_1 = \frac{p_1 f_1}{p_1 - f_1} = \frac{(\frac{1}{4} \times 10^6)10x}{(\frac{1}{4} \times 10^6) - 10x} = 10x \frac{1}{1 - [10x/(\frac{1}{4} - 10^6)]}$$

$$\cong 10x\left(1 + \frac{10x}{4 \times 10^6}\right),$$

$$q_1 = 10x + (4x^2 \times 10^{-4}) \text{ miles} \qquad \text{(real)},$$

$$Y_1 = -\frac{q_1}{p_1} = -\frac{10x}{\frac{1}{4} \times 10^6}\left(\frac{1}{4} \times 10^4\right) = -\frac{x}{10} \text{ miles} \qquad \text{(inverted)},$$

$$p_2 = \left(f_2 - \frac{4x^2}{10^4}\right) \qquad \text{(image is just inside } f_2\text{)},$$

$$p_2 = x\left[1 - \frac{4x}{10^4}\right] \qquad \text{(real)},$$

$$q_2 = \frac{p_2 f_2}{p_2 - f_2} = \frac{x^2[1 - (4x/10^4)]}{-4x^2/10^4} = -\frac{10^4}{4} \qquad \text{(virtual)},$$

$$Y_{\text{im}} = -\frac{-10^4/4}{x}\left(\frac{-x}{10}\right) = -\frac{10^3}{4} \text{ miles} \qquad \text{(inverted)}.$$

Thus, $Y_{\text{im}} < Y_{\text{obj}}$ by a factor of 10. However, the image is closer than the object by a factor of 100, so it subtends an *angle ten times bigger*, which means that it *appears* bigger.

(d) For the opera glass:

$$q_1 = \frac{p_1 f_1}{p_1 - f_1} = \frac{50(0.1)}{50 - 0.1}\,\text{m} = 0.1\left(1 + \frac{0.1}{50}\right) = 0.1 + 0.0002 \qquad \text{(real)},$$

$$Y' = -\frac{0.1}{50}2 = -0.004\ \text{m} \qquad \text{(inverted)},$$

$$p_2 = q_1 - L = 0.05 + 0.0002 \qquad (\text{virtual, so } p_2 = -|p_2|),$$

$$q_2 = \frac{(-0.05 - 0.0002)(-0.05)}{(-0.05 - 0.0002) - (-0.05)}\,\text{m} = \frac{0.0025}{-0.0002}\,\text{m} = -12.5\ \text{m} \ \ \text{(virtual)},$$

$$Y_{\text{im}} = -\frac{-12.5}{-0.05} \cdot \frac{(-0.1)}{50} \cdot 2\text{m} = +\frac{0.1}{0.1} = 1\ \text{m} \qquad \text{(erect)}.$$

The image, 1 m high at 12.5 m distance subtends an *angle* eight times as big as the object, which is 2 m high at 50 m distance. So the angular magnification is +8 and the image appears bigger.

1.8 Isotropic means the same in all directions. Every point of the object radiates isotropically. Only part of the power from each point is intercepted by the lens, but the part which is intercepted will be focused on the corresponding point of the image. Consider a single point radiating power (δP) isotropically, as shown.

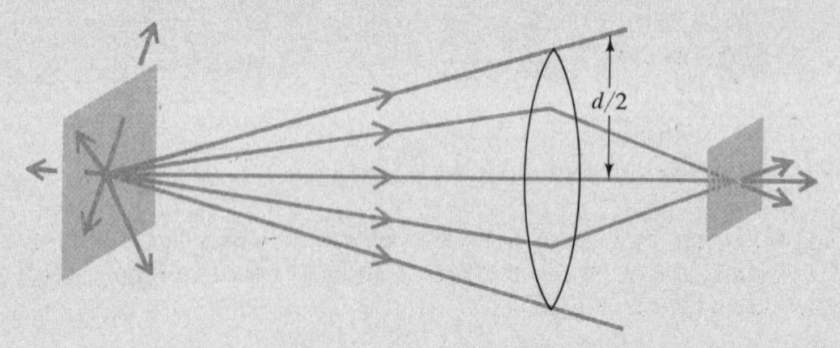

Draw a sphere of radius p as shown. Its surface area is $4\pi p^2$. The area of the lens is $\pi(d/2)^2$. Therefore the fraction of (δP) which goes through the lens is

$$\frac{\pi(d/2)^2}{4\pi p^2} = \frac{d^2}{16p^2}(\delta P).$$

This is true for every point of the object, so the *total* power going through the lens to the image is $Pd^2/16p^2$.

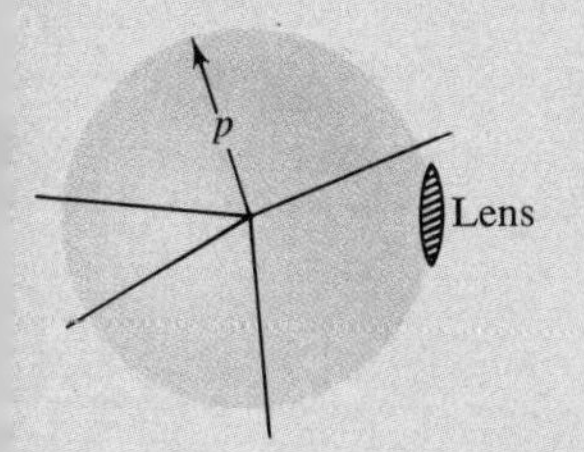

(a)

$$P' = P \cdot \frac{d^2}{16p^2}.$$

(b) Now suppose the object is a square of side ℓ (if it is not, we can divide it into small squares): $A = \ell \times \ell$. How big is the image? The side of the image is $\ell = M\ell = -(q/p)\ell$. So the area of the image is

$$A' = \ell' \times \ell' = \left(\frac{q}{p}\right)^2 A.$$

Thus the intensity at the image is

$$I' = \frac{P'}{A'} = \frac{P(d^2/16p^2)}{A(q^2/p^2)} = \left(\frac{P}{A}\right)\frac{d^2}{16q^2} = I\frac{d^2}{16q^2}.$$

Now

$$q = \frac{pf}{p+f} = f\left(\frac{1}{1+(f/p)}\right).$$

So

$$I' = I \cdot \frac{d^2}{16f^2}\left(1 + \frac{f}{p}\right)^2.$$

(c) Suppose we let $p = 10f$. Then

$$I' = \frac{I}{16} \cdot \left(\frac{d}{f}\right)^2 \left[1 + \frac{1}{10}\right]^2 = 1.21\,\frac{I}{16}\left(\frac{d}{f}\right)^2.$$

This is within 25 percent of the value that we get by ignoring the term in f/p:

$$I' \simeq \frac{I}{16}\left(\frac{d}{f}\right)^2.$$

This is what we do when we calibrate a light meter and use a camera according to f-number (f/d). Clearly, a correction is necessary for close-up work, but for general work, p is always much more than $10f$.

1.10 Image 1: $p = q$, inverted, same size, That is, this image lands right on top of the object. It is not magnified, but it is inverted.

Image 2: There is a virtual image in the plane mirror, a distance f behind it, and erect. This acts as a real object for mirror C: $p_2 = 4f$; so $q_2 = (4/3)f$. Thus the image is closer to C than the first, is one-third the size of the original, and is inverted. Image 2 is accompanied by another real image, at the same place and of the same size, but inverted. This comes from image 1, as reflected in P and then C. All subsequent images will have such companions.

Image 3: Image 2 is also reflected in P, giving a virtual image at $-(5/3)f$ relative to P, which is a real object for C at $p_3 = (14/3)f$. Then $q_2 = (14/11)f$. This means that the image is real and magnified by $-3/11$. This inverts the previous orientation, leaving this image erect with respect to the original and one-eleventh the size of the original object.

All subsequent real images are progressively smaller and progressively closer to the focal point of the curved mirror. This is just what we should have expected, since successive reflections in the plane mirror appear to be farther away.

1.11 In all cases use the single-spherical surface equation:

$$\frac{n_{\text{inc}}}{p} + \frac{n_{\text{em}}}{q} = \frac{n_{\text{em}} - n_{\text{inc}}}{R}.$$

(a)

$$R = \infty \qquad q = -\frac{n_{\text{em}}}{n_{\text{inc}}} p = -\frac{1.0}{4/3} 10 \text{ cm} = -7.5 \text{ cm}.$$

Thus the observer thinks the fish is closer to the glass than it is really, by 2.5 cm. The image of the fish is, of course, virtual.

(b) $R = +5$ cm (center of curvature is in the bubble, which is where light from the fish is emerging).

$$\frac{4/3}{20} + \frac{1}{q} = \frac{1 - 4/3}{5}.$$

(Notice that the 3 cm to the fly has *nothing* to do with this.)

$q = -7.5$ cm (virtual).

(The fly thinks the fish is nearer *him* by 12.5 cm.)

(c)

$R = -5$ cm.

(Light from fly is going in other direction.)

$$\frac{1}{3}+\frac{4/3}{q}+\frac{4/3-1}{-5}.$$

($q=-3\frac{1}{3}$ cm virtual. The fish sees the fly $\frac{1}{3}$ cm farther away.)

(d) Use the image from (c) as an object: $R=\infty$.

$$\frac{4/3}{33\frac{1}{3}}+\frac{1}{q}=0, \qquad q=-25 \text{ cm}.$$

2.1 We can rewrite the expression as

$$y\,(x,\,t)=\exp\left[-a\left(z+\sqrt{\frac{b}{a}}\,t\right)^2\right].$$

(a) The wave is traveling toward $-z$, since as t increases, z must get smaller in order to keep the exponent the same.

(b) The wave speed is

$$c=-\sqrt{\frac{b}{a}}=-\frac{1}{4}\text{ cm/sec}.$$

(c) The sketch should be:

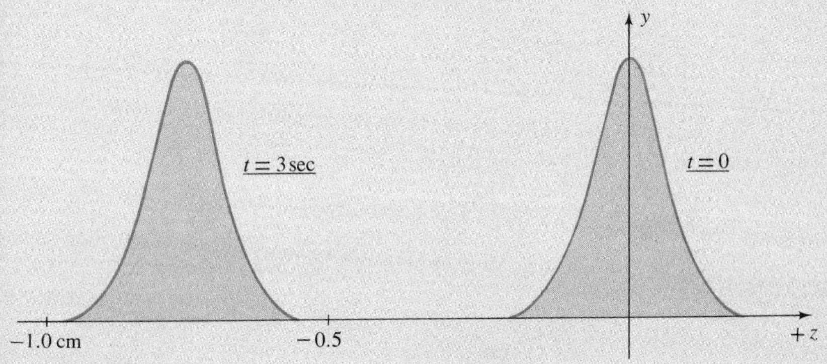

2.3 The slope at the leading edge is $\partial y/\partial x=0.010$. The wave speed is $c=5.0$ m/sec, and the transverse particle speed is $v_y=\partial y/\partial t$. This merely involves differentiations like those in the early part of the chapter: $y=y(x-ct)$; we know that

$$\frac{\partial y}{\partial x}=\frac{\partial y}{\partial(x-ct)}\frac{\partial(x-ct)}{\partial x}=\frac{\partial y}{\partial(x-ct)}.$$

We wish to know

$$\frac{\partial y}{\partial t} = \frac{\partial y}{\partial(x-ct)} \frac{\partial(x-ct)}{\partial t} = -c \frac{\partial y}{\partial(x-ct)}.$$

This means that

$$\frac{\partial y}{\partial x} = -\frac{1}{c}\frac{\partial y}{\partial t}.$$

Therefore

$$v_y = -c\frac{\partial y}{\partial x} = -5.0(\text{m/sec}) \cdot 0.010 = -0.05 \text{ m/sec}.$$

The particle is moving downward, according to the negative sign. This agrees with a picture of this sort:

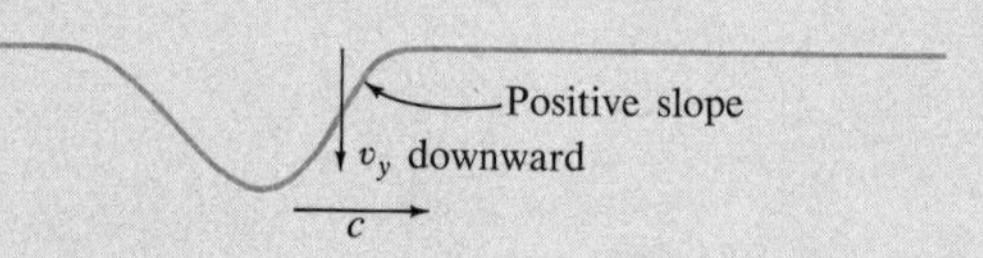

2.6 (a) The equation is of the form

$$y(x, t) = A \sin\left[2\pi\left(\frac{x}{\lambda} + \frac{t}{\tau}\right) + \phi\right].$$

The amplitude A is 5 cm, as read directly from the drawings. The wavelength is 16 beads, which equals 32 cm, since they are 2 cm apart when the string is stretched along the x axis, and have the same separation in the x direction when the wave is present. By comparing the two pictures, we see that the wave has moved 4 cm to the right in 2 sec, so $c = +2$ cm/sec, and $\tau = \lambda/c = 16$ sec. The motion toward $+x$ means that we select the negative sign in the equation:

$$y(0, 2 \text{ sec}) = 0 = A \sin\left[2\pi\left(\frac{0}{\lambda} - \frac{2 \text{ sec}}{16 \text{ sec}}\right) + \phi\right] = A \sin\left(-\frac{\pi}{4} + \phi\right).$$

So $\sin[\phi - (\pi/4)] = 0$. This means $\phi = \pi/4$ or $5\pi/4$; to decide, we look at the slope at $x = 0, t = 2$ sec.

$$\text{Slope} \quad \frac{\partial y}{\partial x} = \frac{A2\pi}{\lambda} \cos\left[2\pi\left(\frac{x}{\lambda} - \frac{t}{\tau}\right) + \phi\right].$$

$$\text{Slope} \quad (0, 2\text{ sec}) = \frac{2\pi A}{\lambda}\cos\left[-\frac{\pi}{4}+\phi\right],$$

and we know this is positive. If $\phi = \pi/4$, slope $(0, 2) = +A\lambda/2\pi$. If $\phi = 5\pi/4$, slope $(0, 2) = -2\pi A/\lambda$, so $\phi = \pi/4$ is the correct choice.

Another way to find the correct value of ϕ is to find the values of x and t at a crest: $y(x_c, t_c) = +A$; so

$$\sin\left[2\pi\left(\frac{x_c}{\lambda}-\frac{t_c}{\tau}\right)+\phi\right] = +1,$$

or

$$\phi + 2\pi\left(\frac{x_c}{\lambda}-\frac{t_c}{\tau}\right) = \frac{\pi}{2}.$$

In this case, (x_c, t_c) can be $(4 \times 2\text{ cm}, 2\text{ sec})$, so

$$\phi + 2\pi\left(\frac{8}{32}-\frac{2}{16}\right) = \frac{\pi}{2} \qquad \text{or} \qquad \phi = \frac{\pi}{2}-\frac{\pi}{4} = \frac{\pi}{4}.$$

This is usually a simpler method.

So the solution for (a) is

$$y(x, t) = (5\text{ cm})\sin\left[2\pi\left(\frac{x}{32\text{ cm}}-\frac{t}{16\text{ sec}}\right)+\frac{\pi}{4}\right],$$

or

$$y(x, t) = (5\text{ cm})\sin\left(\frac{\pi}{16\text{ cm}}\left[x - 2\frac{\text{cm}}{\text{sec}}t\right]+\frac{\pi}{4}\right).$$

(b) At $t = 12$ sec, the square bead has $y = 0$ and is moving upward. To show this, plug in $x = 36$ cm, $t = 12$ sec. Then

$$y(36, 12) = 5\text{ cm}\sin\left(2\pi\cdot\frac{36}{32}-2\pi\cdot\frac{12}{16}+\frac{\pi}{4}\right) = 5\text{ cm}\sin(\pi) = 0,$$

$$\frac{\partial y}{\partial t} = 5\text{ cm}\left(\frac{-2\pi}{16}\text{ sec}\right)\cos(\pi) = +\frac{5\pi}{8}\text{ cm/sec} = v_y.$$

$$\text{Momentum} = p_y = mv_y = 10\text{ g}\cdot\frac{5\pi}{8}\text{ cm/sec} = \frac{50\pi}{8}\text{ g cm/sec}.$$

Note that $p_x = 0$, since the bead never moves in the x direction.

$$\text{Energy} = \text{KE} = \frac{1}{2}mv_y^2 = 5\text{ g}\left(\frac{25\pi^2}{64}\right)\text{ cm}^2/\text{sec}^2 = \frac{125\,\pi^2}{64}\text{ ergs}.$$

(c) Force =

$$ma(x, t) = m\frac{\partial^2 y}{\partial t^2} = -mA(2\pi\nu)^2 \sin(\quad) = -m(2\pi\nu)^2 y(x, t),$$

or

$$F = -10\text{ g}\left(\frac{2\pi}{16}\text{ sec}\right)^2 (5\text{ cm}) \sin\left[2\pi\left(\frac{x}{32\text{ cm}} - \frac{t}{16\text{ sec}}\right) + \frac{\pi}{4}\right],$$

and

$$F(x, t) = F(x - ct) = -\frac{25\pi^2}{32}\left(\frac{\text{g cm}}{\text{sec}^2}\right) \sin\left[2\pi\left(\frac{x}{32\text{ cm}} - \frac{t}{16\text{ sec}}\right) + \frac{\pi}{4}\right].$$

Then, putting in the values for the square bead,

$$F(36\text{ cm}, t) = -\left(\frac{25\pi^2}{32}\right) \cos\left(-\frac{\pi t}{8\text{ sec}}\right)\text{ dynes.}$$

The graph looks like the accompanying illustration.

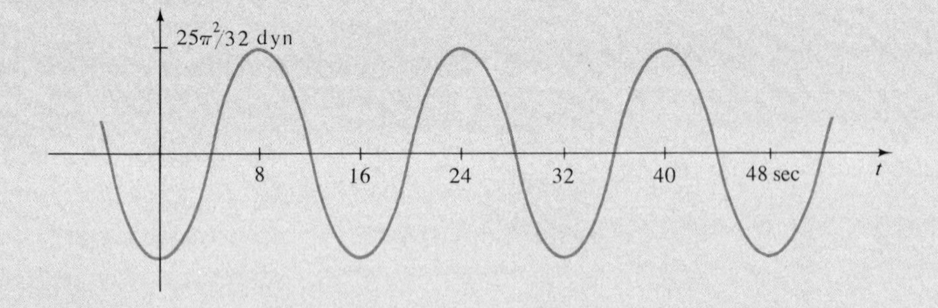

(d) This part depends on material in Appendix B. Since

$$c = 2\text{ cm/sec}, \qquad \rho = 10\text{ g/2 cm}, \qquad c = \sqrt{\frac{T}{\rho}}.$$

Therefore, $T = 20$ dynes.

3.2 The graph of these functions appears on the opposite page.

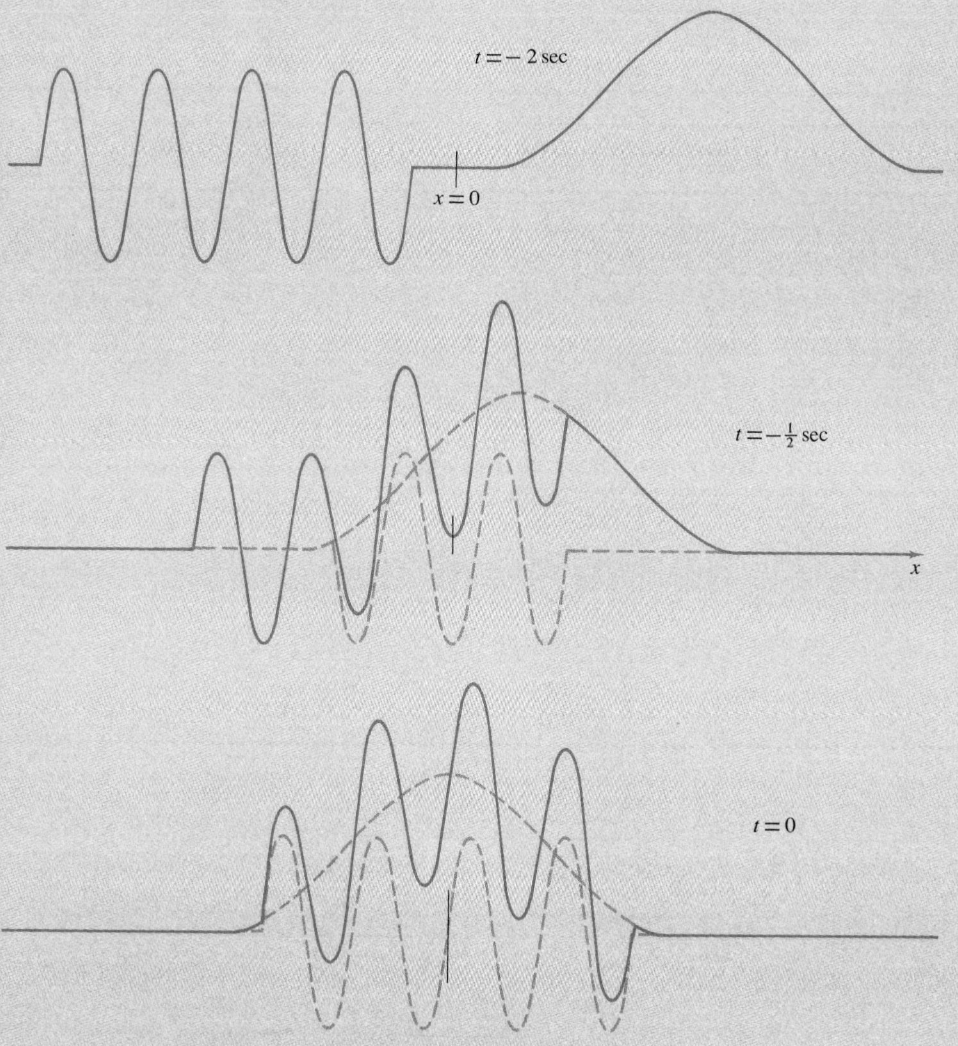

3.4 (a) The individual pressures are

$$p_1(\lambda, 0) = p_0 \cos\left[\frac{2\pi}{\lambda}\lambda + \frac{\pi}{4}\right] = p_0 \cos\frac{\pi}{4},$$

and

$$p_2(\lambda, 0) = \frac{1}{2}p_0 \sin\left[\frac{2\pi}{\lambda}\lambda + \frac{\pi}{4}\right] = \frac{1}{2}p_0 \sin\frac{\pi}{4}.$$

Their sum is

$$p_0 \cos\frac{\pi}{4} + \frac{1}{2} p_0 \sin\frac{\pi}{4} = \frac{\frac{3}{2} p_0}{2}.$$

(b) Again,

$$p_1\left(\lambda, \frac{\tau}{3}\right) = p_0 \cos\left[\frac{2\pi}{\lambda}\lambda - 2\pi\frac{\tau/3}{\tau} + \frac{\pi}{4}\right] = p_0 \cos\left(\frac{\pi}{4} - \frac{2\pi}{3}\right),$$

$$p_2\left(\lambda, \frac{\tau}{3}\right) = \frac{1}{2} p_0 \sin\left[\frac{2\pi\lambda}{\lambda} - 2\pi\frac{\tau/3}{\tau} + \frac{\pi}{4}\right] = \frac{1}{2} p_0 \sin\left(\frac{\pi}{4} - \frac{2\pi}{3}\right).$$

Thus

$$p_T = p_0 \cos\left(-\frac{5\pi}{12}\right) + \frac{1}{2} p_0 \sin\left(-\frac{5\pi}{12}\right) = -0.224 p_0.$$

(c) $z = \lambda$ means that we add or subtract 2π, so this expression can be dropped. [$\sin(\theta + 2\pi) = \sin\theta$.] One approach is to let

$$p_1(\lambda, t) = p_0 \cos\left(-2\pi\nu t + \frac{\pi}{4}\right)$$

$$= p_0 \sin\left(2\pi\nu t + \frac{\pi}{4}\right)$$

and

$$p_2(\lambda, t) = \frac{1}{2} p_0 \sin\left(-2\pi\nu t + \frac{\pi}{4}\right).$$

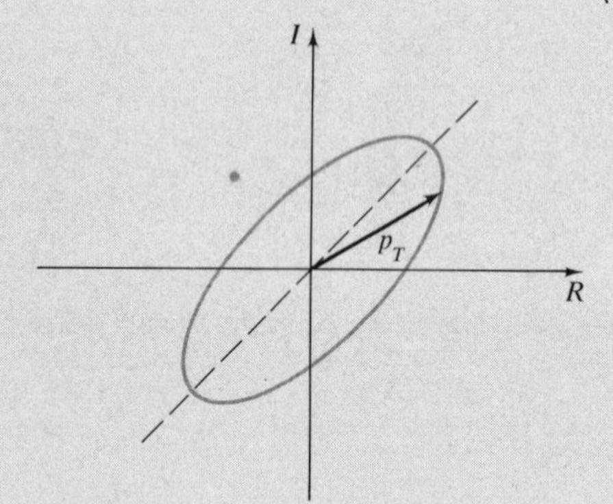

The two phasors p_1 and p_2 rotate in different directions as time increases. They are lined up when they are along $\pi/4$ and $(\pi/4) + \pi$. They oppose when $ct = \pi/2$: The resultant phasor traces out an ellipse, with semi-major axis $= \frac{3}{2}p_0$, semi-minor axis $= \frac{1}{2}p_0$, with axes at 45 degrees to I and R. This phasor rotates at rate $2\pi/\tau$, in the counterclockwise direction. Its projection on the I axis is the quantity that represents p_T.

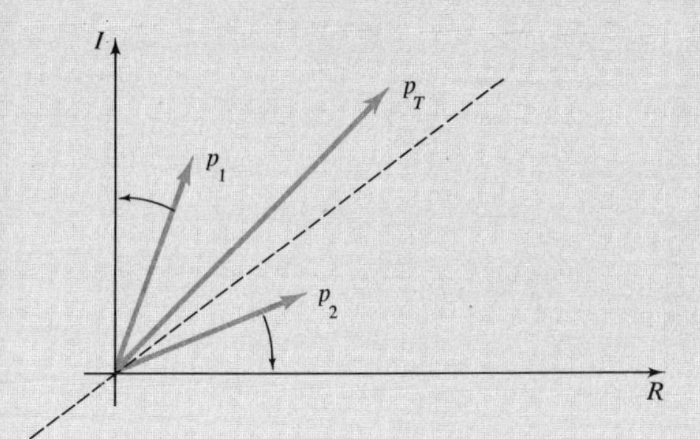

An alternate (and simpler) reduction of the trigonometry is to use

$$p_1 = \sin\left(\frac{3\pi}{4} - 2\pi\nu t\right) \qquad \text{and} \qquad p_2 = \frac{1}{2}\sin\left(\frac{\pi}{4} - 2\pi\nu t\right).$$

These give a constant-length resultant which traces out a circle. The *component along I* is the same.

3.6 The waves (wavelength λ, speed c) are reflected, and then form a standing wave. Since the wall is smooth, this is like the free end of a

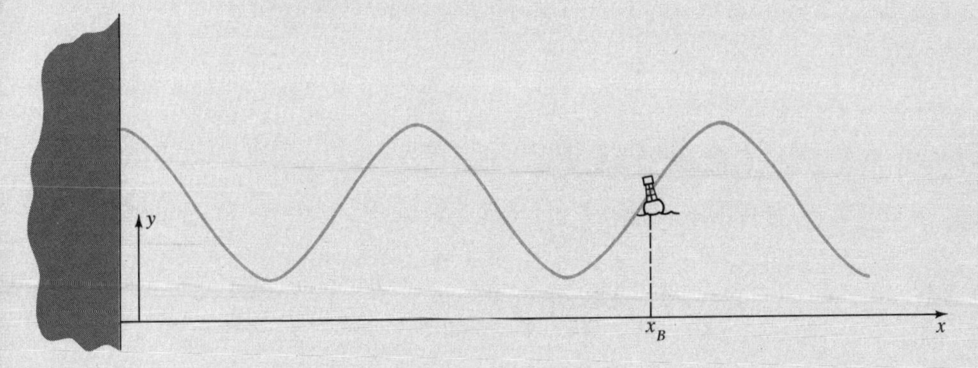

string, so the wall is at an antinode. Let $x = 0$ be at the wall and positive x out to sea. The buoy is at $x = x_B$. The amplitude at point x is

$$y = A\sin\frac{2\pi}{\lambda}(x - ct) + A\sin\frac{2\pi}{\lambda}(-x - ct) = 2A\sin 2\pi\nu t\cos 2\pi\frac{x}{\lambda}.$$

(This does leave the standing-wave antinode at $x = 0$.) The buoy stays on the surface of the water, so its vertical position (relative to the calm sea surface) is also

$$y_B = 2A \sin 2\pi\nu t \cos 2\pi \frac{x_B}{\lambda}.$$

Its velocity is

$$v_B = 4\pi\nu A \cos 2\pi\nu t \cos 2\pi \frac{x_B}{\lambda} \qquad \textit{vertically.}$$

Acceleration is

$$a_B = -8\pi^2\nu^2 A \sin 2\pi\nu t \cos 2\pi \frac{x_B}{\lambda} \qquad \textit{vertically.}$$

Then energy is

$$\mathscr{E} = \frac{1}{2} mv^2 + mgy$$

$$= \frac{m}{2}(16\pi^2\nu^2 A^2) \cos^2 2\pi\nu t \cos^2 \frac{2\pi x_B}{\lambda} + mg(2A) \sin 2\nu t \cos \frac{2\pi x_B}{\lambda},$$

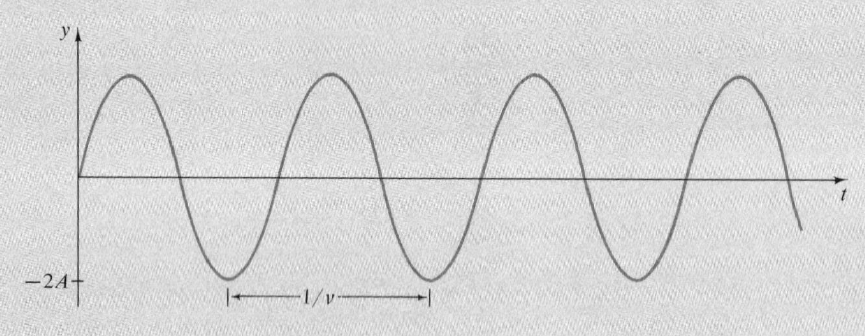

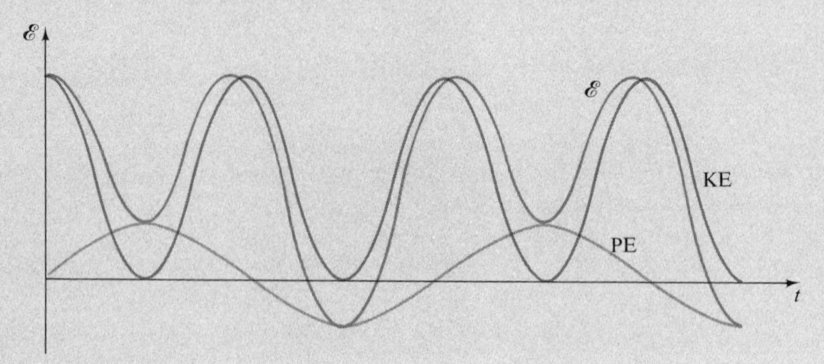

and

$$\mathbf{p} = \hat{\mathbf{j}} 4\pi\nu A m \cos 2\pi\nu t \cos 2\pi \frac{x_B}{\lambda}.$$

3.8 The "frequency" with which you see fence 1 at 20 km/hr is

$$\nu_1 = \frac{c_1}{\lambda_1} = \left(\frac{100 \text{ mm}}{20 \text{ km/hr}}\right)^{-1} = \left[\frac{100 \text{ mm}/(2 \times 10 \text{ mm})}{3600 \text{ sec}}\right]^{-1}.$$

For the second fence, a similar calculation yields 54.5 sec^{-1}. The beat frequency is then 1.1 sec^{-1}. Keeping in mind that the car "hears" beats as loud when crests coincide *and* when troughs coincide, note that the frequency of *not* seeing through the fences occurs midway *between* these two coincidences, and so it will also be *at* the beat frequency.

Another approach to this problem is to inspect the stationary patterns: Look for the distance between coincidences. We can write the position of the boards of fence 1 as: $x_1 = m_1 \cdot 100$ mm, where m_1 is an integer. For fence 2 we can write $x_2 = m_2 \cdot 102$ mm. Coincidences occur when $x_1 = x_2$; that is, when $m_1/m_2 = 102/100 = 1.02$. Since both m_1 and m_2 are integers, this situation occurs when $m_1 = 51, 102, 153$, etc. Consequently, the *distance* between beats is $x_1(51) = 51 \times 100$ mm $= 5.1$ m, and for the car traveling at 20 km/hr,

$$\nu_{\text{beat}} = \frac{2 \times 10^4}{3.6 \times 10^3 \times 5.1} \text{ sec}^{-1}.$$

where $\nu_{\text{beat}} = 1.1 \text{ sec}^{-1}$.

This line of argument will be useful later when we find coincidences among fringes in interference patterns. A more familiar example is shown in the accompanying sketch.

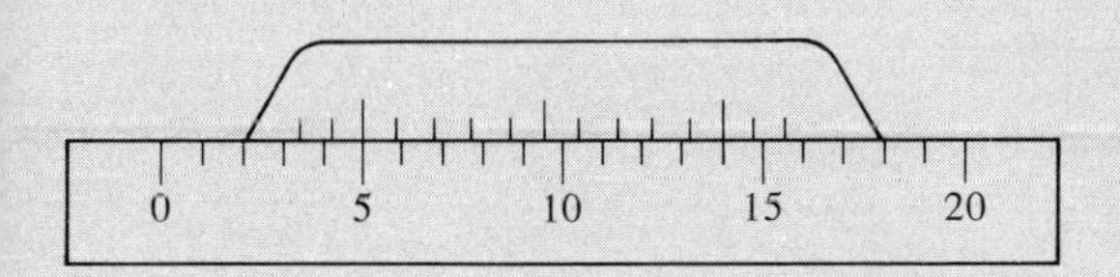

3.10 First we note that since the strings are identical, the wave speed is the same on each. So $\lambda_1 \nu_1 = \lambda_2 \nu_2 = c$. This means that $\lambda_2 = 0.7\lambda_1$. The two strings are shown in the diagram.

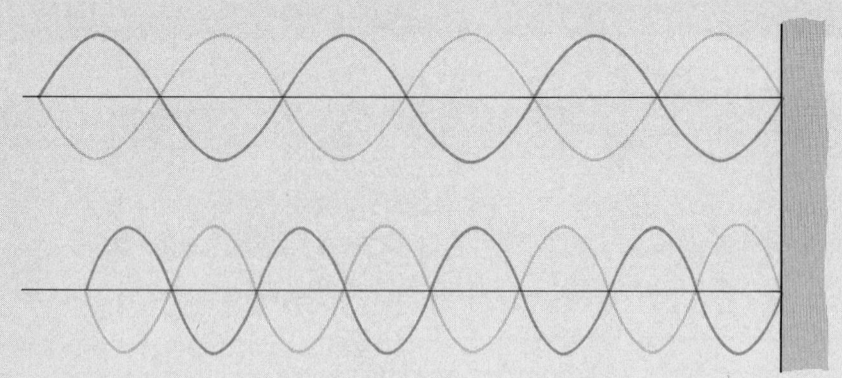

It is usually sufficient to know the positions of the nodes and antinodes, and this problem is a case in point. Naturally, there is a node at the wall, so nodes occur at $x = n(\lambda/2)$, where n is an integer. Antinodes are at $x = (\lambda/2) + (\lambda/4)$.

(a) The point $x = 1.75\lambda_1$ is at an antinode on string 1. So the kinetic-energy density at that point is

$$\frac{K\mathscr{E}}{\text{unit length}} = \frac{1}{2}\rho(2\pi\nu)^2 A^2 \sin^2 2\pi\nu t.$$

(b) Here, $x = 1.75\lambda_1 = 2.50\lambda_2$ is at a node on string 2. The string does not move at a node, so $K\mathscr{E}/\text{unit length} = 0$.

(c) To find the answer to this part, we can extend the drawing and observe, or we can solve the moiré pattern analytically. Since this kind of analysis will arise often, we should do the latter. We note that nodes on string 1 occur at $x_1 = n_1\lambda_1/2$, and those on string 2 at $x_2 = n_2\lambda_2/2$. For a coincidence, $x_1 = x_2$, or $n_1\lambda_1 = n_2\lambda_2$. This means that

$$n_1\lambda_1 = n_2\frac{7}{10}\lambda_1 \qquad \text{or} \qquad \frac{n_1}{n_2} = \frac{7}{10}.$$

Since n_1 and n_2 are integers, we can find possible values for them: $n_2 = 0, 10, 20, 30, \ldots$; $n_1 = 0, 7, 14, \ldots$, and the positions of the coincidences is found from either:

$$x_{\text{coincid}} = \frac{n_1\lambda_1}{2} = 0, 3.5\lambda_1, 7\lambda_1, 10.5\lambda_1 \cdots$$

or

$$x_{\text{coincid}} = \frac{n_2\lambda_2}{2} = 0.5\lambda_2, 10\lambda_2, 15\lambda_2, \cdots,$$

giving the same value of x.

3.12 Wave velocity $= c = \nu\lambda = \sqrt{g\lambda/2\pi}$.

(a) Group velocity $= \partial\nu/[\partial(1/\lambda)]$. Then

$$\nu = \sqrt{\frac{g}{2\pi\lambda}}, \qquad \text{and} \qquad \frac{\partial\nu}{\partial(1/\lambda)} = \sqrt{\frac{g}{2\pi}}\,\frac{1}{2}\left(\frac{1}{\lambda}\right)^{-1/2},$$

so

$$v_g = \frac{1}{2}\sqrt{\frac{g\lambda}{2\pi}} = \frac{1}{2}\,c.$$

(b)

$$y_1 = A\sin\left[2\pi\,\frac{x}{\lambda_1} - 2\pi\nu_1 t\right],$$

$$y_2 = A\sin\left[2\pi\,\frac{x}{\lambda_2} - 2\pi\nu_2 t\right].$$

Therefore,

$$y_T = y_1 + y_2$$

$$= 2A\sin\left[\pi x\left(\frac{1}{\lambda_1} + \frac{1}{\lambda_2}\right) - \pi(\nu_1 + \nu_2)t\right]\cos\left[\pi x\left(\frac{1}{\lambda_2} - \frac{1}{\lambda_1}\right) - \pi(\nu_1 - \nu_2)t\right].$$

The apparent velocity of the carrier wave is taken from the first factor:

$$v = \frac{\text{coefficient of } t}{\text{coefficient of } x} = \frac{\nu_1 + \nu_2}{(1/\lambda_1) + (1/\lambda_2)} \cong \nu_1\lambda_1 \quad \text{or} \quad \nu_2\lambda_2.$$

The velocity associated with the beat pattern is extracted in the same way from the second factor:

$$v_g = \frac{\text{coefficient of } t}{\text{coefficient of } x} = \frac{\nu_1 - \nu_2}{(1/\lambda_1) - (1/\lambda_2)}.$$

We know λ_1 and λ_2, and

$$\nu_1 = \sqrt{\frac{g}{2\pi}\,\frac{1}{\lambda_1}},$$

$$\nu_2 = \sqrt{\frac{g}{2\pi}\,\frac{1}{\lambda_2}}.$$

Therefore

$$v = \nu_1\lambda_1 = 3.16\sqrt{\frac{g}{2\pi}} \qquad \text{or} \qquad v = \nu_2\lambda_2 = 3.32\sqrt{\frac{g}{2\pi}},$$

and

$$v_g = 1.66\sqrt{\frac{g}{2\pi}}.$$

Thus the "group" velocity calculated this way is in fact just about one-half the wave velocity.

3.14 (a) The pulse is described as

$$F(x,t) = \Delta\nu \sin\theta_0(x, t)\frac{\sin\Delta\theta(x,t)}{\Delta\theta(x,t)},$$

where

$$\theta_0 = 2\pi\left(\frac{x}{\lambda_0} - \nu_0 t\right) \qquad \text{and} \qquad v_w = \lambda_0\nu_0,$$

and

$$\Delta\theta = \pi\left[x\,\Delta\left(\frac{1}{\lambda}\right) - t\cdot\Delta\nu\right] \qquad \text{with } v_g = \frac{\Delta\nu}{\Delta(1/\lambda)}.$$

(b) We can define the spread in position as the width of the pulse halfway between its first zeros: At $t = 0$, this is $\Delta x = 1/\Delta[(1/\lambda)]$. So

$$\Delta x\,\Delta\nu = \frac{\Delta\nu}{\Delta(1/\lambda)} = v_g.$$

For light in vacuum, $v_g = v_w$; therefore

$$\Delta x\,\Delta\nu = c \qquad \text{or} \qquad \Delta x\,\Delta p = h,$$

as suggested. This is in some circumstances a reasonable model for a photon.

4.2 Let us put the metal at $z = 0$ and let the incident wave be

$$\mathbf{E}_i(z, t) = \hat{\mathbf{i}}E_0 \sin\left[\frac{2\pi}{\lambda}(z + ct)\right],$$

$$\mathbf{B}_i(z, t) = -\hat{\mathbf{j}}\frac{E_0}{c}\sin\left[\frac{2\pi}{\lambda}(z + ct)\right].$$

We can get the proper reflected wave from

$$\mathbf{E}_r(z, t) = +\hat{\mathbf{i}}E_0 \sin\left[\frac{2\pi}{\lambda}(z - ct)\right].$$

To verify that this yields the correct node:

$$\mathbf{E}_T(z, t) = \mathbf{E}_i + \mathbf{E}_r = \hat{\imath} 2E_0 \sin \frac{2\pi}{\lambda} z \cos 2\pi\nu t,$$

which is always zero at $z = 0$.

We find $\mathbf{B}_r(z, t)$ in the usual way: its direction is such as to give $\mathbf{E}_r \times \mathbf{B}_r$ in the $+\hat{\mathbf{k}}$ direction, and it is in phase with $\mathbf{E}_r$.

$$\mathbf{B}_r(z, t) = +\hat{\jmath} \frac{E_0}{c} \sin\left[\frac{2\pi}{\lambda}(z - ct)\right].$$

Then

$$\mathbf{B}_T(z,t) = \mathbf{B}_i + \mathbf{B}_r = \hat{\jmath} 2 \frac{E_0}{c} \sin 2\pi\nu t \cos \frac{2\pi}{\lambda} z.$$

This has an *antinode* at $z = 0$! Thus the E and B standing waves are 90 degrees out of phase, even though both component traveling waves have E and B in phase.

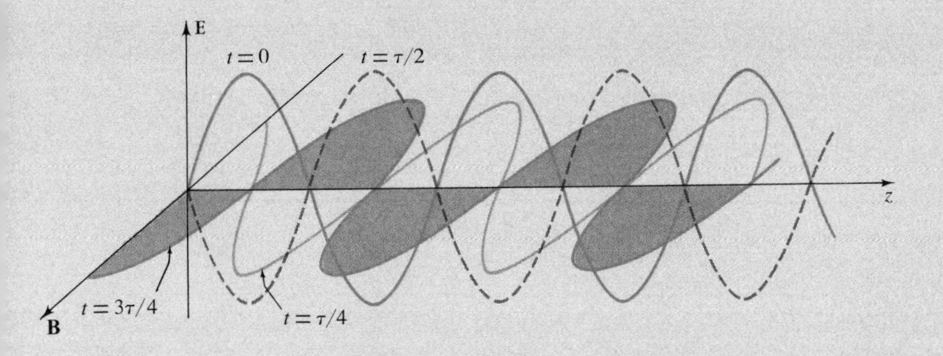

4.4 The wave shown must have the form $\mathbf{E}(z, t) = \hat{\imath} E_0 \cos[(2\pi/\lambda)(z + ct)]$, where $E_0 = 5$ V/m, and $\lambda = 7$ m.

(a) **B** must be at right angles to **E** and the propagation direction. It must be in phase with **E**, and $B_0 = E_0/c$.

$$\mathbf{B}(z, t) = \pm\hat{\jmath} \frac{E_0}{c} \cos\left[\frac{2\pi}{\lambda}(z \pm ct)\right].$$

To verify this, note that if we choose the upper sign, the wave propagates toward $-z$: $\mathbf{E} \times \mathbf{B}$ is in the direction $(\hat{\imath}) \times (-\hat{\jmath}) = -\hat{\mathbf{k}}$. The lower sign signifies propagation toward $+z$.

(b) The momentum density has the magnitude $p/V = \mathscr{E}/c$. Therefore

$$\frac{p}{V} = \frac{\left\{\frac{1}{2}\epsilon_0 E_0^2 \cos^2\left[\frac{2\pi}{\lambda}(z+ct)\right] + \frac{1}{2}\frac{B_0^2}{\mu_0}\cos^2\left[\frac{2\pi}{\lambda}(z+ct)\right]\right\}}{c}$$

$$= \frac{\epsilon_0}{c} E_0^2 \cos^2\left[\frac{2\pi}{\lambda}(z+ct)\right],$$

where we have used $\epsilon_0 \mu_0 = 1/c^2$.

The direction of $\mathbf{p}$ is the propagation direction; so,

$$\frac{\mathbf{p}}{V} = \mp \hat{\mathbf{k}} \frac{\epsilon_0}{c} E_0^2 \cos^2\left[\frac{2\pi}{\lambda}(z+ct)\right], \qquad E_0 = 5 \text{ V/m}, \quad \lambda = 7 \text{ m}.$$

4.7 This problem can be solved by momentum conservation: the initial momentum of the flashlight is MV_i. The final momentum of the flashlight is MV_f. The initial momentum of the light beam is 0. The final momentum of the light beam is $(\mathscr{E}/c) \times$ (volume of beam). Here $\mathscr{E}$ is the energy per unit volume of the beam. But $\mathscr{E} \times$ volume = total energy emitted in time $T = P \cdot T$, so we do not have to find the volume after all! Therefore:

Total momentum initially is $p_T = MV_i + 0$,

Total momentum finally is $p_T = MV_f + \dfrac{PT}{c}$.

These must be equal; so $MV_i = MV_f + (PT/c)$, or

$$V_f = V_i - \frac{PT}{Mc} = V_i - \frac{(1 \text{ W})(3600 \text{ sec})}{M(3 \times 10^8 \text{ m/sec})}.$$

The minus sign merely signifies that the flashlight and the light beam move in opposite directions. Incidentally, the volume of the beam is its cross-sectional area times cT.

4.8 We will take visible light to mean light of wavelength 5000 Å $= 5 \times 10^{-7}$ m (greenish color). The momentum of the photon before collision is $p = h/\lambda = 1.3 \times 10^{-27}$ g/sec. Its energy is $U = 4 \times 10^{-19}$ J. After the collision, the electron has momentum mv, and its energy is $\frac{1}{2}mv^2$, while the reflected photon has momentum $p' = h/\lambda'$ and energy $U' = p'c$. Both quantities are conserved; so

$$p = -p' + mv,$$

$$U = U' + \tfrac{1}{2}mv^2;$$

or

$$p = p' + \frac{1}{2}\frac{mv^2}{c}.$$

So,

$$v = c\left\{\sqrt{1 + \frac{4h}{mc\lambda}} - 1\right\} \simeq \frac{2h}{m\lambda}\left(1 - \frac{h}{m\lambda c}\right).$$

The electron gains nearly $2p$ in momentum. We then use this to find that

$$U' = U\left(1 - \frac{2h}{mc\lambda}\right) = \frac{h}{\lambda'} \cdot c.$$

Solving for λ', we find

$$\lambda' = \frac{\lambda}{1 - (2h/mc\lambda)} \simeq \lambda\left(1 + \frac{2h}{mc\lambda}\right)$$

or $\lambda' \simeq \lambda + (2h/mc) = \lambda + 0.05$ Å, which means almost no change in wavelength. It *is* still visible.

4.11 This is an exercise in deciding who is observing what. It could be solved without reference to the Doppler effect. First pick a frame of reference: the river is the simplest choice. Then the skindiver (S) will see no Doppler shift; he moves with the medium and sees $\nu = \nu_0$. Formally,

$$\nu_s = \nu_0 \frac{c - 0}{c - 0} = \nu_0,$$

where c is the wave speed, 20 km/hr. v_{ob} is the skindiver's velocity (=0). No source velocity enters this calculation. The fisherman (F) is moving north (relative to the river water) at 10 km/hr, so the waves pass him *less* frequently by a factor of 2. Again, formally

$$\nu_F = \nu_0 \frac{c - v_F}{c} = \nu_0 \frac{+20 - 10}{+20} = \frac{\nu_0}{2}.$$

The balloonist (B) is stationary relative to the *wind*, so he goes faster than the waves and measures a frequency of $-\nu_0$:

$$\nu_B = \nu_0 \frac{+20 - 40}{+20} = -\nu_0,$$

the negative sign meaning that he *measures* the waves as going southward relative to him. So the ratios are

$$\nu_S : \nu_F : \nu_B = 2 : 1 : -2.$$

The same result may be found by using $v_F = 0$ and adding in the velocity of the river, as was done with the wind velocity in the example in the text. Also note that if there were a source velocity, v_{sc}, it would multiply *all* the frequencies by $1/[(1 + v_{sc})/c]$ or by $1/[(1 + v_{sc})/(c + v_m)]$ so that the result would not change.

4.12 The Doppler shift for light is given by

$$\nu' = \nu\sqrt{\frac{c \pm v}{c \mp v}}.$$

We may choose the direction of v. Let the mirror be receding so that it absorbs the light as if it were an observer moving away at speed v, and re-emits as if it were a source moving away at speed v. In both cases, the frequency *decreases*; so we want the lower sign in the preceding equation.

$$\nu_{\text{abs}} = \nu\sqrt{\frac{c - v}{c + v}}.$$

$$\nu_{\text{emit}} = \nu_{\text{abs}}\sqrt{\frac{c - v}{c + v}} = \nu\,\frac{c - v}{c + v}.$$

The beat frequency is $\nu_{\text{beat}} = \nu - \nu_{\text{emit}}$:

$$\nu_{\text{beat}} = \nu\left[1 - \frac{c - v}{c + v}\right] = \nu\,\frac{2v}{c + v} = \frac{c}{\lambda}\,\frac{2v}{c + v}.$$

$$\nu_{\text{beat}} \simeq \frac{2v}{\lambda} \qquad \text{when } v \ll c.$$

5.1 $\mathbf{E} \times \mathbf{B}$ must be along $+\hat{\mathbf{k}}$, and $\mathbf{E}$ is in phase with $\mathbf{B}$; so

$$\mathbf{B}(z, t) = (\hat{\mathbf{j}} + (-\hat{\mathbf{i}}))\,\frac{E_0}{c}\sin\left[\frac{2\pi}{\lambda}z - 2\pi\nu t + \frac{7\pi}{6}\right].$$

This means that

$$\mathbf{B}(0, 0) = (-\hat{\mathbf{i}} + \hat{\mathbf{j}})\,\frac{E_0}{c}\sin\frac{7\pi}{6}.$$

(Remember that $\hat{\mathbf{i}} \times \hat{\mathbf{j}} = \hat{\mathbf{k}}$ and that $\hat{\mathbf{j}} \times -\hat{\mathbf{i}} = \hat{\mathbf{k}}$.)

5.2

$$\mathbf{p}(0, t) = \frac{\hat{\mathbf{k}}\mathscr{E}(0, t)}{c} = \frac{\hat{\mathbf{k}}\epsilon_0\mathbf{E}(0, t)\cdot\mathbf{E}(0, t)}{c} = \hat{\mathbf{k}}\,\frac{\epsilon_0}{c}\,2E_0^{\,2}\sin^2\left[\frac{7\pi}{6} - 2\pi\nu t\right].$$

Averaged over time, this becomes

$$\langle \mathbf{p} \rangle_{av} = \hat{\mathbf{k}} \frac{\epsilon_0}{c} E_0{}^2.$$

(Note that $(\hat{\imath} + \hat{\jmath}) \cdot (\hat{\imath} + \hat{\jmath}) = 2$.)

5.4 $$\mathbf{E}\left(-\frac{\lambda}{2}, 0\right) = (\hat{\imath} + \hat{\jmath})E_0 \sin\left[-\pi - 2\pi\nu \cdot 0 + \frac{7\pi}{6}\right]$$
$$= (\hat{\imath} + \hat{\jmath})E_0 \sin\left[\frac{\pi}{6}\right].$$

Phasor E_x Phasor E_y Vector **E**

5.6 At $z = L/2$, the wave is inside the material, and the index of refraction for this polarization and frequency is $n_x = 4$. Thus, $v_w = c/4$ and $\lambda = c/4\nu$. $\lambda = 0.125 \mu\text{m} = 1250$ Å. The wavelength in vacuum here is 5000 Å.

5.8 This material should be transparent for the x polarization up to about 6×10^{14} Hz, and for the y polarization up to about 8×10^{14} Hz. In the region $6\text{–}8 \times 10^{14}$ Hz, the material is transparent for the y polarization, but absorbs the x polarized radiation; so it is dichroic in that range.

5.10 The group velocity is

$$v_g = \frac{\partial \nu}{\partial(1/\lambda)} \quad \text{and} \quad \frac{1}{\lambda} = \frac{n\nu}{c}.$$

Now n depends on the frequency; thus, over a short range, we can say

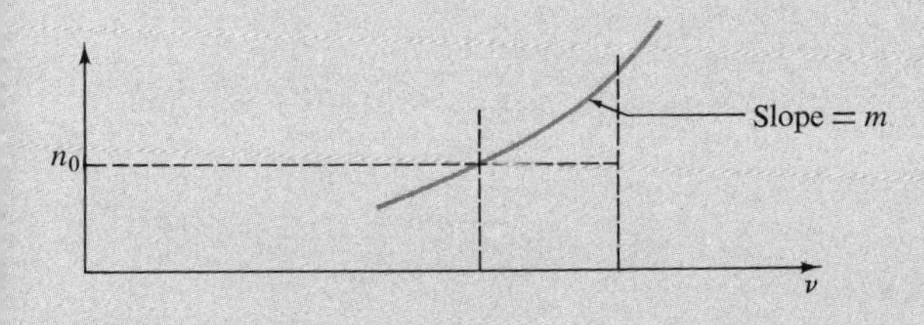

that $n = n_0(1 + m\nu)$, where m is the *slope* of the curve for $n(\nu)$, and n_0 is the value of n at the lower frequency end. This allows us to say that $1/\lambda = (n_0/c)(\nu + m\nu^2)$. If we differentiate both sides of this with respect to $1/\lambda$, we get

$$1 = \frac{n_0}{c}\left[\frac{\partial\nu}{\partial(1/\lambda)} + 2m\frac{\partial\nu}{\partial(1/\lambda)}\nu\right] = \frac{n_0}{c}(1 + 2m\nu)v_g,$$

or

$$v_g = \frac{v_w}{1 + 2m\nu}.$$

So, $v_g < v_w$ whenever m is positive. This means for the x polarization that $v_g < v_w$ for all frequencies below 6×10^{14} Hz. When the slope turns negative, $v_g > v_w$. All we have done to get this result is to approximate a curve with a straight line for a short distance; hence the result is quite generally correct. An alternative method for extracting the same result involves finding v_g explicitly by setting $\nu = c/n\lambda$ in $v_g = \partial\nu/[\partial(1/\lambda)]$. Thus

$$v_g = \frac{\partial}{\partial(1/\lambda)}\left(\frac{c}{n}\cdot\frac{1}{\lambda}\right) = \frac{c}{n} + \frac{c}{\lambda}\frac{\partial(1/n)}{\partial(1/\lambda)} = v_w + \frac{c}{\lambda}\left(\frac{1}{n^2}\right)\frac{\partial n}{\partial(1/\lambda)}$$

$$= v_w\left(1 - \frac{v_g}{\lambda n}\frac{\partial n}{\partial\nu}\right),$$

since

$$\frac{\partial}{\partial(1/\lambda)} = \frac{\partial\nu}{\partial(1/\lambda)}\frac{\partial}{\partial\nu} = v_g\frac{\partial}{\partial\nu}.$$

Then, rearranging,

$$v_g = \frac{v_w}{1 + (\nu/n)(\partial n/\partial\nu)}.$$

5.12 The main difference between this problem and problem 2 is that now $\epsilon \neq \epsilon_0$, and the proper value of ϵ differs with polarization. Remember that $\epsilon_0\mu_0 = 1/c^2$, and $\epsilon\mu = 1/v_w^2$; therefore if we take $\mu = \mu_0$, then $\epsilon = n^2\epsilon_0$. So,

$$\mathbf{p} = \hat{\mathbf{k}}\left[\frac{\epsilon_0}{c}n_x^2E_x^2 + \frac{\epsilon_0}{c}n_y^2E_y^2\right].$$

Putting in numbers and the arguments of E_x and E_y, we find

$$\mathbf{p} = \frac{\epsilon_0}{c}E_0^2\hat{\mathbf{k}}\left\{n_x^2\sin^2\left[\frac{\pi L}{\lambda_0}n_x - 2\pi\nu t + \frac{7\pi}{6}\right]\right.$$

$$\left. + n_y^2\sin^2\left[\frac{\pi L}{\lambda_0}n_y - 2\pi\nu t + \frac{7\pi}{6}\right]\right\}.$$

The average is

$$\langle \mathbf{p} \rangle_{av} = \frac{1}{2}\frac{\epsilon_0}{c} E_0{}^2\hat{\mathbf{k}}(n_x{}^2 + n_y{}^2).$$

5.14 $E_x(L, 0) = E_0 \sin\left[\dfrac{2\pi L}{\lambda_0} n_x + \dfrac{7\pi}{6}\right],$

$$E_y(L, 0) = E_0 \sin\left[\frac{2\pi L}{\lambda_0} n_y + \frac{7\pi}{6}\right].$$

$$\Delta\phi = \frac{2\pi L}{\lambda_0}(n_x - n_y),$$

where

$$\lambda = 0.5\mu\text{m} = 5000\ \text{Å}, \qquad n_x - n_y = 2.$$

Therefore

$$\Delta\phi = \frac{2\pi}{2\mu\text{m}} \cdot \frac{5}{32\mu\text{m}} \cdot 2 = \frac{8\pi}{32} \cdot 5 = \frac{5\pi}{4}.$$

This is a quarter-wave plate.

6.1 (a)

$$E_x = E_0 \sin\left[2\pi\left(\frac{z}{\lambda} - \nu t\right)\right],$$

$$E_y = E_0 \cos\left[2\pi\left(\frac{z}{\lambda} - \nu t\right)\right].$$

These are actual components of a real vector, *not* phasors, so we want to construct the resultant at various times and also some convenient place like $z = 0$.

t	0	$\tau/4$	$\tau/2$	$3\tau/4$	$\tau/8$
E_x	0	$-E_0$	0	E_0	$-E_0/\sqrt{2}$
E_y	E_0	0	$-E_0$	0	$+E_0/\sqrt{2}$

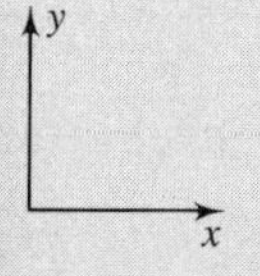

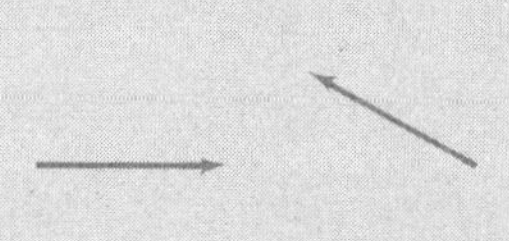

The resultant vector is simply a vector of length E, rotating counterclockwise. The polarization is circular.

(b)

$$E_x = E_0 \sin\left[2\pi\left(\frac{z}{\lambda} + \nu t\right)\right]$$

$$E_y = E_0 \sin\left[2\pi\left(\frac{z}{\lambda} + \nu t - \frac{1}{8}\right)\right]$$

Again we consider the point $z = 0$ and follow the resultant as time progresses.

t	0	$\tau/8$	$2\tau/8$	$3\tau/8$	$4\tau/8$	$5\tau/8$	$6\tau/8$
E_x	0	$+E_0 \sin\frac{\pi}{4}$	E_0	$E_0 \sin\frac{\pi}{4}$	0	$-E_0 \sin\frac{\pi}{4}$	$-E_0$
E_y	$-E_0 \sin\frac{\pi}{4}$	0	$+E_0 \sin\frac{\pi}{4}$	E_0	$E_0 \sin\frac{\pi}{4}$	0	$-E_0 \sin\frac{\pi}{4}$

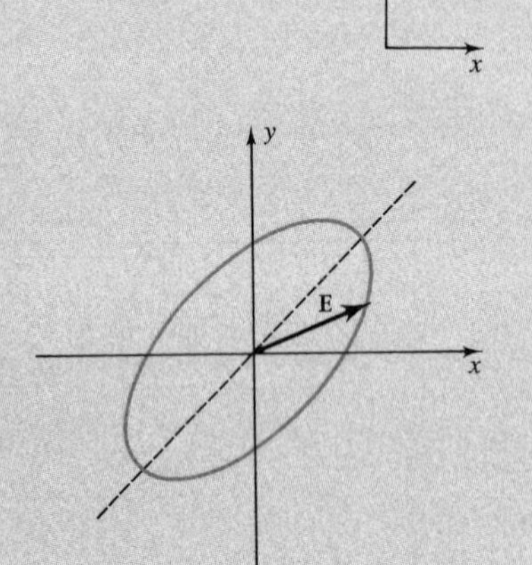

The locus of the end of $\mathbf{E}_{\text{Resultant}}$ is an ellipse at 45 degrees to the axes.

To find the semi-major axis, look at the time when $E_x = E_y$: this is halfway between $t = 0$ and $t = \tau/8 : t = \tau/16$. Hence

$$E_x = E_0 \sin\left(\frac{\pi}{8}\right), \qquad E_y = E_0 \sin\left(\frac{-\pi}{8}\right);$$

$$E_R = E_0 \sin\left(\frac{\pi}{8}\right)\sqrt{2} = 1.31 E_0.$$

For the semi-minor axis, $\pi/2$ further on, $t = \tau/16 + \tau/4 = 5\tau/16$. Then

$$E_x = E_0 \sin\left(\frac{5\pi}{8}\right), \qquad E_y = E \sin\left(\frac{3\pi}{8}\right), \qquad E_R = E_0 \sin\left(\frac{3\pi}{8}\right)\sqrt{2} = 0.542E_0.$$

(c)

$$E_x = E_0 \sin\left[2\pi\left(\frac{z}{\lambda} - \nu t\right)\right],$$

$$E_y = -E \sin\left[2\pi\left(\frac{z}{\lambda} - \nu t\right)\right].$$

As before, $z = 0$ is most convenient place.

t	0	$\tau/8$	$\tau/4$	$3\tau/8$	$4\tau/8$	$5\tau/8$
E_x	0	$E_0 \sin \frac{\pi}{4}$	E_0	$+E_0 \sin \frac{\pi}{4}$	0	$-E_0 \sin \frac{\pi}{4}$
E_y	0	$-E_0 \sin \frac{\pi}{4}$	$-E_0$	$-E_0 \sin \frac{\pi}{4}$	0	$E_0 \sin \frac{\pi}{4}$

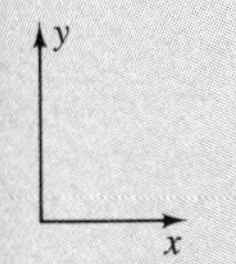

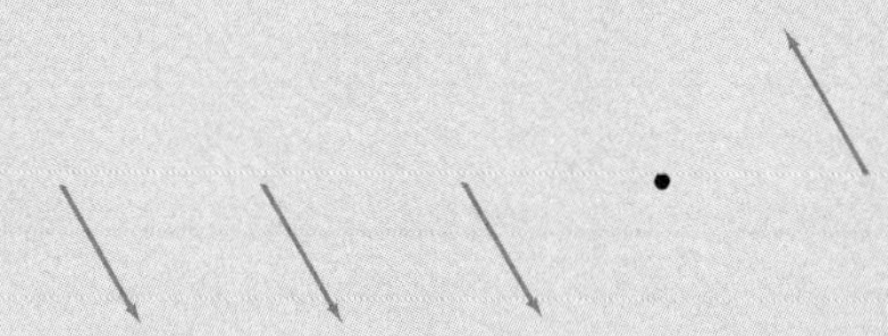

The polarization is thus *linear*, along a line at 45 degrees to the axes. Amplitude is $E_0\sqrt{2}$.

6.3 (a)

$\mathbf{E}_1 = (\hat{\imath}A_1 + \hat{\jmath}B_1) \cos 2\pi\nu t,$

$\mathbf{E}_2 = (\hat{\imath}A_2 + \hat{\jmath}B_2) \cos 2\pi\nu t.$

Again, add components:

$\mathbf{E}_T = \{\hat{\imath}(A + A') + \hat{\jmath}(B + B')\} \cos 2\pi\nu t.$

The point here is that the quantity in { } just specifies a vector, and the time variation is the same for all parts of it. So the vector lies along a line and changes size, but *not* direction. In the case of circular polarization, this is not true, since the x and y components vary differently in time, one being large when the other is small.

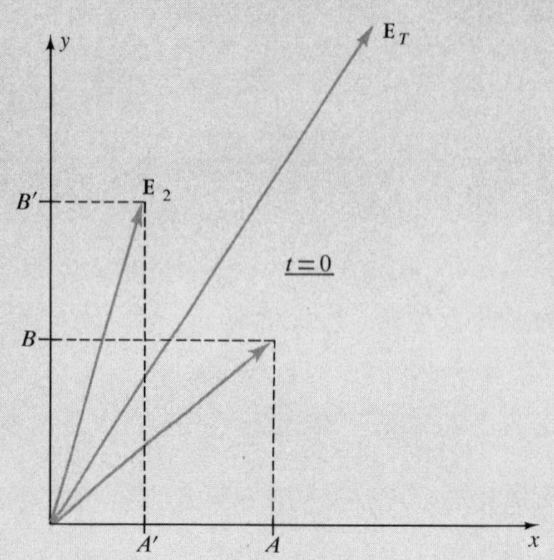

Left: $\quad \mathbf{E}_1 = \hat{\mathbf{i}}E_0 \cos 2\pi\nu t + \hat{\mathbf{j}}E_0 \sin 2\pi\nu t,$

Right: $\quad \mathbf{E}_2 = \hat{\mathbf{i}}E_0 \cos(2\pi\nu t + \alpha) - \hat{\mathbf{j}}E_0 \sin(2\pi\nu t + \alpha).$

$$\mathbf{E}_\mathrm{T} = \hat{\mathbf{i}}E_0\{\cos 2\pi\nu t + \cos(2\pi\nu t + \alpha)\} + \hat{\mathbf{j}}E_0\{\sin 2\pi\nu t - \sin(2\pi\nu t + \alpha)\}$$

$$= 2\hat{\mathbf{i}}E_0 \cos\left(2\pi\nu t + \frac{\alpha}{2}\right) \cos\frac{\alpha}{2} + 2\hat{\mathbf{j}}E_0 \sin\frac{\alpha}{2} \cos\left(2\pi\nu t + \frac{\alpha}{2}\right)$$

$$= 2E_0\left(\hat{\mathbf{i}} \cos\frac{\alpha}{2} + \hat{\mathbf{j}} \sin\frac{\alpha}{2}\right) \cos\left(2\pi\nu t + \frac{\alpha}{2}\right).$$

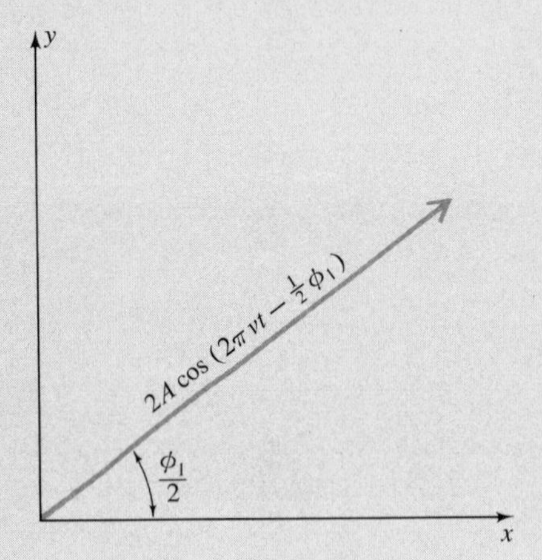

Since the two components are in phase, we can add them in a *vector* diagram: $\mathbf{E}_T$ is a vector of magnitude $2A\cos[2\pi\nu t - (\phi_1/2)]$ at an angle of $\phi_{1/2}$ to the y axis.

Notice that we can see some easy limiting cases in the first two lines: If $\alpha = 0$, the y components cancel out, while the x components add to $\mathbf{E}_T = \hat{\imath}2E_0 \cos 2\pi\nu t$ ($\alpha = \pi$ just multiplies everything by -1). If $\alpha = \pi/2$, we get

$$\mathbf{E}_T = 2E_0(\hat{\imath} + \hat{\jmath})\{\cos 2\pi\nu t + \sin 2\pi\nu t\},$$

linearly polarized at 45 degrees.

(c)

$$\mathbf{E} = \hat{\imath}A \sin 2\pi\nu t + \hat{\jmath}B \cos 2\pi\nu t.$$

First take out a circularly polarized part:

$$\mathbf{E}_0 = \hat{\imath}A \sin 2\pi\nu t + \hat{\jmath}A \cos 2\pi\nu t.$$

What is left is $\mathbf{E}_{\text{lin}} = \hat{\jmath}(B - A)\cos 2\pi\nu t$, a linearly polarized wave. To show that the original wave is elliptically polarized, draw the *vector* at various times:

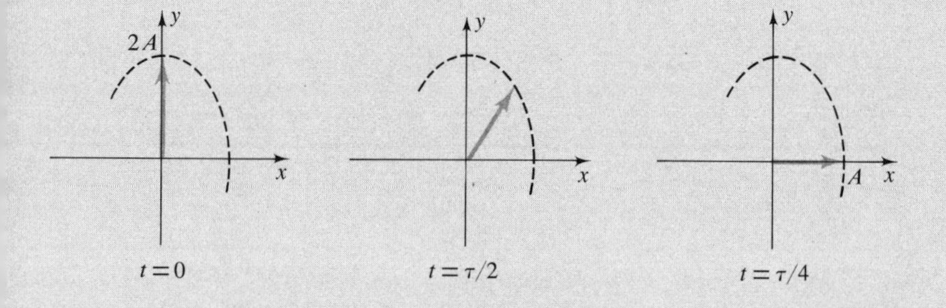

6.5 Intensity before P is I_i. Intensity after P is $I_i/2$ (on average, half the energy is tied up in each polarization).

(a) Intensity after S is

$$\frac{I_i}{2} \cdot \sin^2 45° = \frac{I_i}{4}.$$

Intensity after A is

$$\frac{I_i}{4} \cdot \sin^2 45° = \frac{I_i}{8} = I_f.$$

(b) Intensity after S is still $I_i/2$, but the plane of polarization has been rotated through $2 \times (45°) = 90°$ so that it is now lined up with the analyzer. So the intensity after A is $I_i/2 = I_f$.

(c) Again, S does not change the intensity, but rather turns the polarization into circular, with intensity $I_i/2$. Then the analyzer transmits the half of this which is lined up with it. Remember, we can decompose circularly polarized light into two linearly polarized components along *any* mutually perpendicular axes. So $I_f = I_i/4$.

6.7 The first device, a polarizer, reduces intensity to $I_0/2$ and polarizes along, say, the $\theta = 0°$ line. The next device, a half-wave plate, has its axis at $\theta = \pi/4N$, and so rotates the polarization through twice this angle to $\theta = \pi/2N$. It absorbs no energy. The third device has its axis along $\theta = (\pi/4N) \cdot 2 = \pi/2N$, so it has no effect. The next half-wave plate rotates the light another $\pi/2N$, and so forth. Every other half-wave plate has no effect. Those N which do something rotate it through $(\pi/2N)N = \pi/2$ so that it goes through the final polarizer without loss. The final intensity is $I_f = I_0/2$. To check this, try with $N = 1$ (one half-wave plate) and $N = 2$.

6.9 We refer to Figure 6.10 and consider the four components before and after the $\lambda/4$ plate. Before, $E_{\text{fast}} = E \cos\theta$, $E_{\text{slow}} = E \sin\theta$. Then

$$E_{\text{fast}\,\parallel} = E\cos\theta\cos\theta, \qquad E_{\text{fast}\,\perp} = -E\cos\theta\sin\theta,$$

$$E_{\text{slow}\,\parallel} = E\sin\theta\sin\theta, \qquad E_{\text{slow}\,\perp} = E\sin\theta\cos\theta,$$

and

$$E = E_0 \sin 2\pi\left(\frac{z}{\lambda} - \nu t\right) = E_0 \sin\alpha(t).$$

After the quarter-wave plate, each slow component has been slowed by $\pi/2$ relative to the fast components. So

$$E'_{\text{fast}\,\parallel} = E_{\text{fast}\,\parallel}; \qquad E'_{\text{fast}\,\perp} = E_{\text{fast}\,\perp},$$

$$E'_{\text{slow}\,\parallel} = E'\sin^2\theta, \qquad E'_{\text{slow}\,\perp} = E'\sin\theta\cos\theta,$$

where $E' = E_0 \cos\alpha(t)$.

Now recombine: $E_\parallel = E\cos^2\theta + E'\sin^2\theta$. This is the component that the analyzer eliminates. Also

$$E_\perp = -E\cos\theta\sin\theta + E'\cos\theta\sin\theta,$$

which the analyzer passes, and this means

$$E_\perp = E_0\cos\theta\sin\theta[\cos\alpha(t) - \sin\alpha(t)],$$

$$I_\perp = I_0 \cos^2\theta \sin^2\theta[\cos^2\alpha(t) + \sin^2\alpha(t) - 2\cos\alpha(t)\sin\alpha(t)].$$

So the average intensity is $I_\perp = I_0 \cos^2\theta \sin^2\theta$.

6.11 The *Handbook of Chemistry and Physics** gives the specific rotation of sucrose as +66.4 degrees for 10 cm of dilute solution. The half-wave plate must therefore be set with its axis at −33.2 degrees, to restore the polarization to its original orientation. Here the plus sign means right-handed rotation. Which axis of the half-wave plate we measure from makes no difference.

6.14 We have used the instantaneous values, unlike the averages of the linear momentum method. This is possible because the geometry shows the frictional effects (as θ). Now let us say that

$$\mathbf{F} = qE_0[\hat{\mathbf{i}}\cos\phi(t) + \hat{\mathbf{j}}\sin\phi(t)],$$

$$\mathbf{r} = -r[\hat{\mathbf{i}}\cos(\phi(t)+\delta) + \hat{\mathbf{j}}\sin(\phi(t)+\delta)],$$

and

$$\mathbf{v} = r[\hat{\mathbf{i}}\cdot 2\pi\nu\sin(\phi(t)+\delta) - \hat{\mathbf{j}}2\pi\nu\cos(\phi(t)+\delta)],$$

treating each component independently. Then

$$\langle P_{\text{abs}}\rangle = \frac{1}{\tau}\int_0^\tau \mathbf{F}\cdot\mathbf{v}\,dt$$

$$= \frac{1}{\tau}qE_0 2\pi\nu r\int_0^\tau [\cos\phi\sin(\phi+\delta) - \sin\phi\cos(\phi+\delta)]\,dt,$$

and

$$\langle\boldsymbol{\Gamma}\rangle = \frac{1}{\tau}\int_0^\tau \mathbf{F}\times\mathbf{r}\,dt$$

$$= \frac{1}{\tau}qE_0 r\int_0^\tau [-\hat{\mathbf{k}}\cos\phi\sin(\phi+\delta) + \hat{\mathbf{k}}\sin\phi\cos(\phi+\delta)]\,dt.$$

So

$$\langle P_{\text{abs}}\rangle = \frac{\langle\Gamma\rangle}{2\pi\nu}.$$

There is no particular advantage to this form, unless one has a model for δ.

* "Handbook of Chemistry and Physics", 48th edition (R. C. Weast, ed.). Chem. Rubber Publ. Co., Cleveland, Ohio, 1967.

6.15 The first device absorbs half the power and leaves the wave linearly polarized. The light momentum before (a) was TP/c, and after (a) is $\frac{1}{2}(TP/c)$. So (a) has acquired momentum $\frac{1}{2}(TP/c)$ in the direction of the wave travel. The original light beam had no angular momentum, and the same is true after (a); so (a) acquires no angular momentum. The quarter-wave plate in device (b) absorbs no energy, and therefore has no linear momentum. It does change the polarization from linear to circular. Before (b) the wave had zero angular momentum; afterward it has $\frac{1}{2}(TP/2\pi\nu)$. So the quarter-wave plate acquires the angular momentum, $TP/4\pi\nu$. The absorber, device (c), acquires the linear momentum $TP/2c$ and the angular momentum $TP/4\pi\nu$ from the light. We can then put in numbers to get:

	(a)	(b)	(c)
Final linear momentum:	$\dfrac{5P}{2c}$	0	$\dfrac{5P}{2c}$
Final angular momentum:	0	$-\dfrac{5P\lambda}{4\pi c}$	$+\dfrac{5P\lambda}{4\pi c}$

6.17 Consider first the pulse perpendicular to the plane of the web: This makes the central point C go up and down, starting pulses on all strings. If the initial pulse had amplitude A (and energy $\mathscr{E} = KA^2$) then each outgoing pulse gets one-eighth the energy and thus has amplitude $A/\sqrt{8}$. Detection is the same on all strings. If the pulse is *in* the plane of the web, no pulse is propagated on string S because point C moves along string S and cannot generate a transverse pulse. For string T, point C moves at 45 degrees to the string, and its component across the string does generate a wave: here the amplitude is smaller than that on strings R and U by a factor of cos 45°. The total energy is conserved; so

$$A^2 = 2A_U{}^2 + 4A_T{}^2 = 2A_U{}^2 + 4\frac{A_U{}^2}{2} = 4A_U{}^2.$$

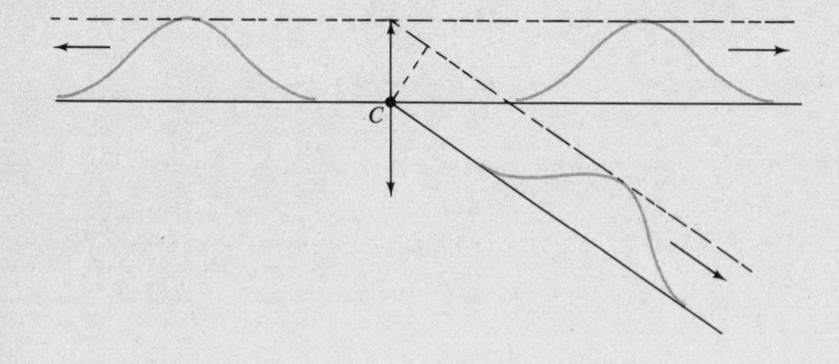

So

$$A_U = \frac{1}{2}A, \qquad A_T = \frac{A}{\sqrt{8}}, \qquad A_s = 0.$$

7.3 Our definition is not of "coherence", but of "coherence time", T_c. The transmitters have constant relative phase while they are on — say for 20 hr. The next day, $\Delta\phi$ may be different. So $\Delta\phi$ is constant for periods of about 20 hr and this is the coherence time. Therefore, the transmitters are "coherent" if measured by a detector of $T_d < 20$ hr and are incoherent otherwise. The important thing to note here is that the coherence depends on the observer. The *coherence time* is the property characteristic of the system, and is independent of the observer.

7.4 The coherence time here is about 0.1 μs, so we can answer the questions as follows:

(a) The eye has a detection time of a few milliseconds; so it will *not* detect interference. The fringes will shift position many times during the time necessary for the eye to perceive them and thus they appear as a uniformly illuminated blur.

(b) The phototube has $T_d = 10^{-9}$ sec; so it will register the fringe positions before they change, thereby detecting interference.

(c) This, like the answer to Problem 7.3, depends on the observer. For visual observation, the answer is *no*; for the fast phototube, the answer is *yes*.

7.6 Sources A and B are identical, hence give a constant fringe pattern: $I_{AB} = \langle (E_A + E_B)^2 \rangle_{av} = I_A \cos^2(\phi/2)$, where ϕ is a constant depending on the source separation. The same is true of C and D. $I_{CD} = I_C \cos^2(\phi'/2)$. The pair may then be regarded as single sources with somewhat complicated intensity distributions, and these (two) sources interfere detectably when $T_d < T_c$. Thus, when $T_d < T_c$,

$$I_T = (\sqrt{I_A}\cos\phi/2 + \sqrt{I_C}\cos\phi'/2)^2 \cdot \cos^2\frac{\phi - \phi'}{2}.$$

When $T_d > T_c$, on the other hand, the (two) sources appear incoherent, and so we observe

$$I_T = I_A \cos^2\frac{\phi}{2} + I_C \cos^2\frac{\phi}{2}.$$

7.8 The observer on the z axis is equidistant from the two sources, so only the relative phases of the sources are of interest.

(a) The phases are the same, so the observer hears the beat note at 50 Hz as well as the carrier at 8.975 kHz.

(b) The frequencies are the same, so when $\phi_A = \phi_B$, the interference is constructive. However, the value of ϕ_B changes every microsecond, so the observer cannot follow it; 10^{-4} sec is his limit. (He hears a buzz.)

(c) This is the same as part (b) except that the change is slow enough for the listener to follow. He hears a tone at 8.975 kHz and a 100-Hz hum. Since the modulation is by a square wave, the frequencies present are more complicated, as can be seen from Appendix H.

8.3 This arrangement, known as Fresnel's double mirrors, allows two plane-wave fronts to overlap. The two reflected half-beams (semicircular in cross section) interfere where they overlap, giving a pattern like that

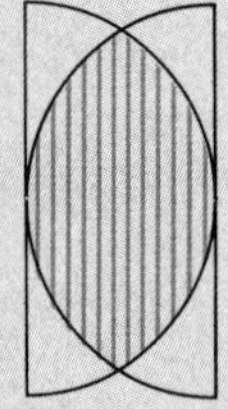

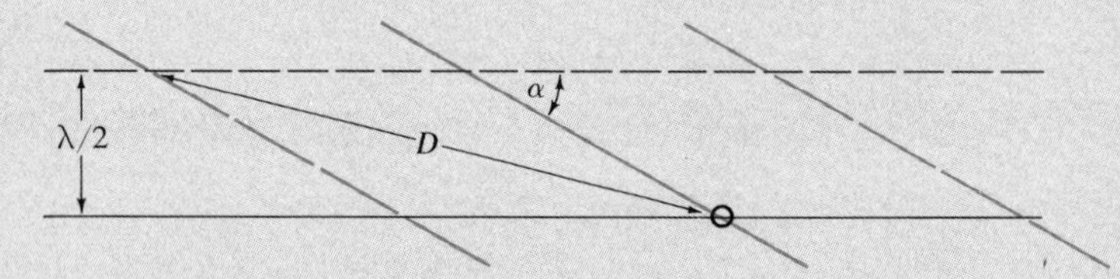

sketched. In the region of overlap, where fringes are formed by the intersection of plane-wave fronts, we may find the fringe spacing in a manner similar to that of Problem 8.13(b). This approach shows that

$$D = \frac{\lambda/2}{\sin(\alpha/2)},$$

where α is the angle between the mirrors. For our small angle, $D = \lambda/\alpha$. We can formulate our problem as a two-source one in order to use the expression we have already developed: The fringes may be thought of as due to two point sources at some large distance L away and separated by the angle α. Then

$$d = 2L \sin\frac{\alpha}{2} \qquad \text{and} \qquad I = I_0 \cos^2\left[\frac{2\pi}{\lambda} L \sin\theta \sin\frac{\alpha}{2}\right]$$

This gives us a fringe spacing of $D = (\lambda/2)/\sin(\alpha/2)$, which agrees with the other result. Since L and θ are arbitrary, we might replace them by a measured position along the detection plane (a film, perhaps): $Y = L \sin\theta$. Then

$$I = I_0 \cos^2\left(\frac{2\pi}{\lambda} Y \sin\frac{\alpha}{2}\right).$$

Notice that the geometry here is the difficulty rather than the concepts of interference. This is why such approximations as that of small angles are so important. A more exact solution contributes nothing to our understanding of the physics.

8.5 First we must decide which sources are coherent. Light from Betelguese falls on the two slits, and they form a pair of coherent sources. The same is true for light from Rigel. But the two stars are not coherent with respect to each other. So we have a two-slit pattern due to each star, centered on the line joining the star and the slits. These two patterns add incoherently (that is, the intensities add). We can do this formally as shown in the figure below.

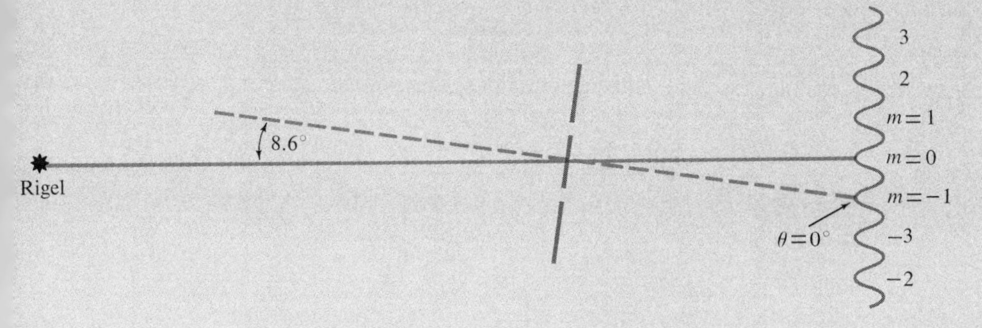

The figure shows that for one star, the slits have an apparent separation of $d\cos(8.6°) = 0.97d$ and that the pattern is centered on $\theta = 8.6°$. Hence we write

$$I_R(\theta) = I_0 \cos^2\left[\frac{\pi}{\lambda}(0.97d)\sin(\theta - 8.6°)\right].$$

Similarly

$$I_B(\theta) = I_0 \cos^2\left[\frac{\pi}{\lambda}(0.97d)\sin(\theta + 8.6°)\right].$$

Now we add the intensities to find the total pattern:

$$I_T(\theta) = I_0 \cos^2\left[\frac{0.97\pi d}{\lambda}\sin(\theta - 8.6°)\right] + I_0 \cos^2\left[\frac{0.97\pi d}{\lambda}\sin(\theta + 8.6°)\right].$$

The student should check for himself that tilting the slit does not shift the position of the $m = 0$ fringe from the star-slit axis. A good approach is to include a phase lag for the upper slit, due to the plane wave's reaching it later, and an extra path length from the lower slit. It is easy to show that these cancel.

8.6 Here we are concerned only with the position of the bright fringes. With no gas in either cell, the central fringe (equal numbers of wavelengths in each path, $m = 0$) falls at the symmetric point P. With gas in A there will be $\ell n/\lambda$ wavelengths in A, but only ℓ/λ in B. So the $m = 0$ fringe must move *toward A* in order to shorten the upper path and lengthen the lower one. Since the central fringe has moved to the previous $m = 20$ position, it must be that cell A now contains 20 more wavelengths than cell B:

$$\frac{\ell n}{\lambda} = \frac{\ell}{\lambda} + 20 \qquad \text{or} \qquad n = 1 + \frac{20\lambda}{\ell}.$$

We can do all this more formally by writing down each wave separately:

$$E_A = E_0 \sin\left(2\pi\nu t + \frac{2\pi}{\lambda}\left(L_A - \ell\right) + \frac{2\pi}{\lambda/n}\ell\right),$$

$$E_B = E_0 \sin\left(2\pi\nu t + \frac{2\pi}{\lambda} L_B\right).$$

After the usual manipulations, we find

$$I_T = I_0 \cos^2\left[\frac{\pi}{\lambda}(L_A - L_B) + \frac{\pi}{\lambda}\ell(n-1)\right].$$

The $m = 0$ fringe occurs at

$$\frac{L_A - L_B}{\lambda} + \frac{\ell}{\lambda}(n-1) = 0.$$

So, as n increases, so must L_B. At the position of the previous $m = 20$ fringe, $L_A - L_B = 20\lambda$, so we are back to the earlier result. (This device is known as a Rayleigh refractometer.)

8.8 In this case, although the light is never polarized, the birefringent material will make a difference. Consider first that component of the light which is polarized along the slit (say, vertically); the path through the half-wave plate over slit B contains, say, N wavelengths. That through slit A then contains $N + \frac{1}{2}$ wavelengths, with the result that the symmetric center point ($L_A = L_B$) is now a dark fringe and the equal time (or equal number of wavelengths, $m = 0$) point has shifted half a fringe toward A.

The horizontal polarization component (H) does not, of course, interfere with the vertical one, since vectors can cancel only if they are collinear. This fringe pattern also shifts, this time half a fringe toward B. The result is, as shown, a restored fringe pattern, with a dark fringe at the center.

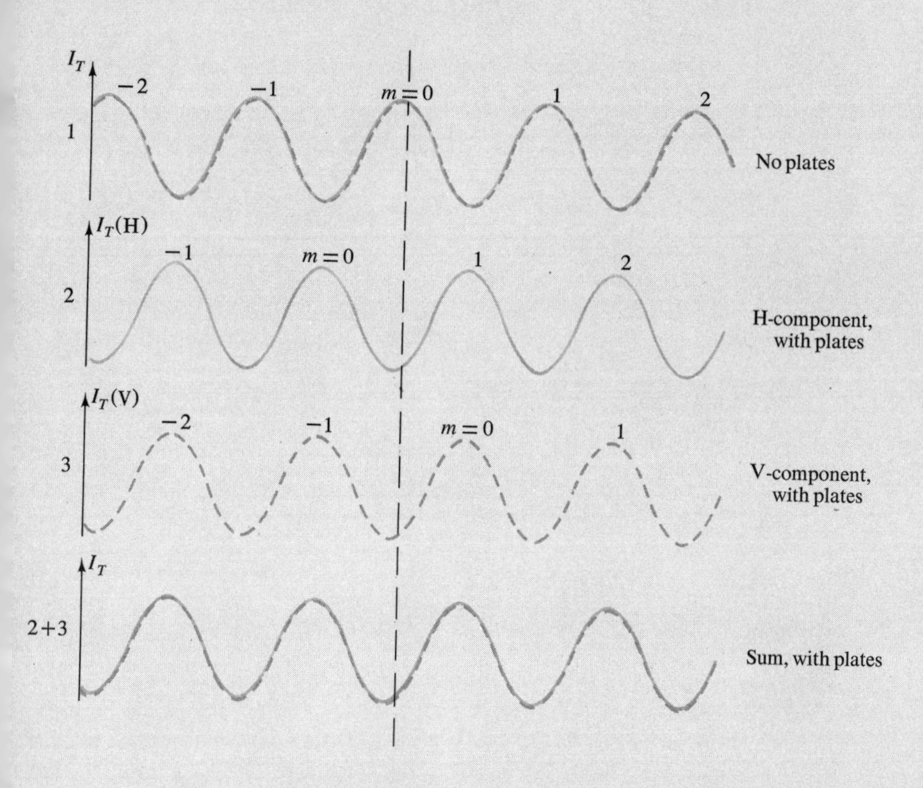

8.12 In part (a) the light coming through the slits is only that with wavelength λ_1. Bright fringes occur at $\sin\theta = m\lambda/d$, so $\sin\theta_9 = 9\lambda_1/d$.

(b) Dark fringes occur at $\sin\theta' = (m + \frac{1}{2})\lambda/d$, and this time the wavelength is λ_2. Hence $\sin\theta_9' = 9\frac{1}{2}\lambda_2/d$.

(c) If $T_d < 1/\Delta\nu$, we have a situation like that described in Problem 8.14. But if the more realistic situation is the case, $T_d > 1/\Delta\nu$, and the sources

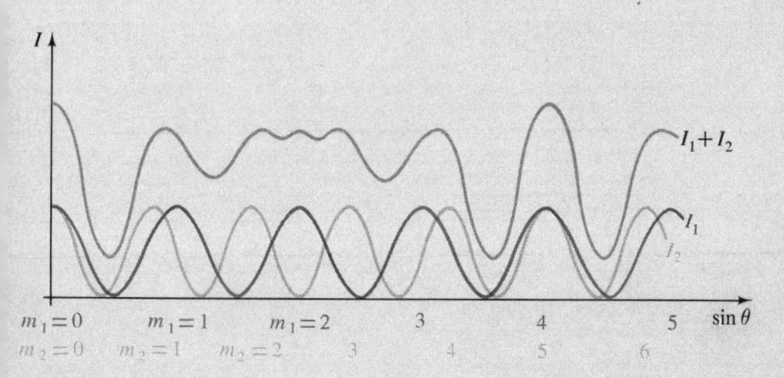

are incoherent, then the intensities add, and $I = 2I_0$, not a function of θ.

(d) Both colors produce interference patterns, and these are both present at the detector. They are incoherent, so their intensities add. We look for the place where bright fringes from both patterns first coincide: this is the point where $d \sin\theta = m_1\lambda_1 = m_2\lambda_2$. Now both m_1 and m_2 are integers, and they differ by a smaller integer, Δm. At the first coincidence, $\Delta m = 1$; at the next, $\Delta m = 2$; and so forth. Therefore

$$\frac{m_1+1}{m_1} = \frac{\lambda_1}{\lambda_2} = \frac{5}{4} \qquad m_1 = 4.$$

$$\sin\theta_4 = \frac{4\lambda_1}{d} = 0.002.$$

8.13 All the wave fronts are plane, since the distances are large. In (a) the fringe at the observer might be bright or dark, depending on the value of d.

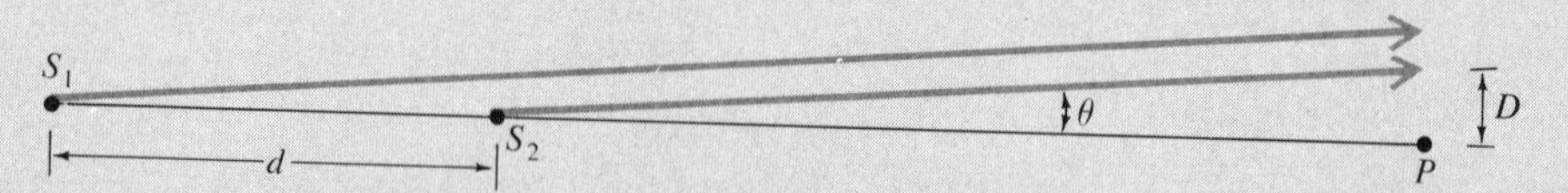

A little to the side of the axis the rays from S_1 and S_2 are nearly parallel, and the next fringe occurs at $d\cos\theta = \lambda$. The actual fringe spacing, in terms of the distance to source 2, is

$$D = L_2 \sin\theta_1 = L_2 \frac{\sqrt{d^2 - \lambda^2}}{d} = \left(1 - \frac{1}{2}\frac{\lambda^2}{d^2}\right) L_2 .$$

An alternative arrangement is that shown in the second figure. Again the fringe at P is not determined, but the spacing is:

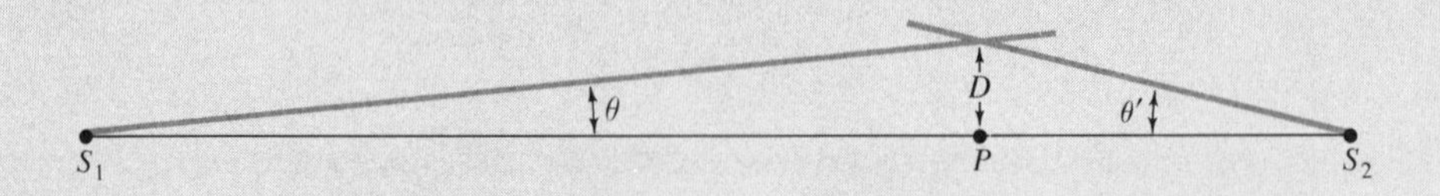

$$L_1 \tan\theta_1 = L_2 \tan\theta_2 \qquad \text{and} \qquad \frac{L_1}{\sin\theta_1} + \frac{L_2}{\sin\theta_1'} = \lambda,$$

where, as usual, the subscript 1 refers to the point one fringe away from the center. Again we take small angles (since the general problem yields the equation of a conic), and write

$$\frac{L_1}{\theta_1} + \frac{L_2}{\theta_1'} = \lambda, \qquad L_1\,\theta_1 = L_2\,\theta_1' = D.$$

$$D = \frac{L_1{}^2 + L_2{}^2}{\lambda}.$$

In (b) we have a single geometric arrangement. The wave fronts intersect as shown, with bright fringes marked by circles. These are easily shown to be separated by $D = \lambda$.

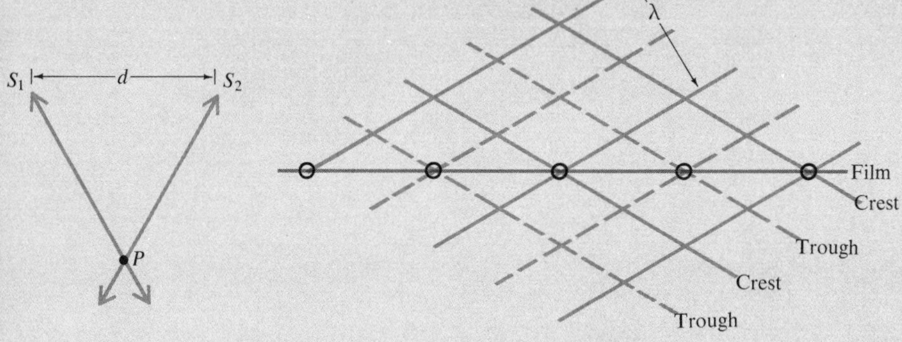

In (c) the wave fronts always intersect at right angles. We can think of the observer here as a strip of film along the circumference. Thus the spacing D is an arc length. From the figure,

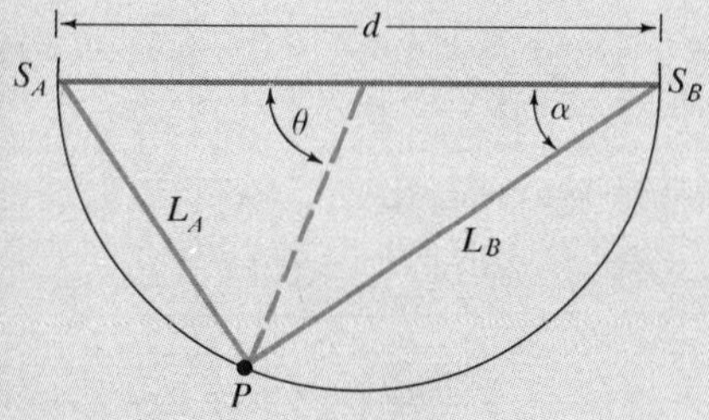

$$L_A = d \sin \alpha = d \sin \frac{\theta}{2} \qquad L_B = d \cos \frac{\theta}{2}.$$

$$L_A - L_B = N\lambda, \qquad L_{A_1} - L_{B_1} = (N+1)\lambda,$$

where the subscript 1 refers to the next fringe. Subtracting,

$$\Delta(L_A - L_B) = \lambda, \qquad \Delta L_A = d\left(\sin \frac{\theta_1}{2} - \sin \frac{\theta}{2}\right) = \frac{d}{2}\,\Delta\theta \cos \frac{\theta}{2}.$$

$$\Delta L_B = -\frac{d}{2}\,\Delta\theta \sin \frac{\theta}{2}.$$

Here $\Delta x \equiv x_1 - x$, and we can find Δ (sin or cos) by trigonometry, using the small-angle approximations, or by differentiation. So

$$\lambda = \frac{1}{2} d\, \Delta\theta \left(\sin\frac{\theta}{2} + \cos\frac{\theta}{2} \right).$$

The fringe spacing is $D = \frac{1}{2} d\, \Delta\theta$, so we can write

$$D = \frac{\lambda}{(\sin \frac{1}{2}\theta + \cos \frac{1}{2}\theta)},$$

which is easily confirmed at $\theta = 0$, π, and $\pi/2$.

Notice that we have used three different approaches to these solutions. In (a) we have found the (small) angle between the $m=0$ and the $m=1$ fringe. In (b) we have measured the actual distance between wave-front intersections in a fixed geometry, and in (c) we have used the path-length difference to arrive at a general expression.

8.14 The problem here is that the frequency of one of the beams will be doppler shifted. We know that $\Delta\nu/\nu = v/c$ for small velocities, so we must expect our interference pattern to shift at the beat frequency $\Delta\nu$. However, the pattern does not simply pulsate like the beating of two sound waves. Rather, as the bright fringes go dark, the dark ones brighten. The same is really true of the acoustic case, but with light we will see both dark and bright fringes at once. The net result is that the fringe pattern appears to move across the field of view, as described by the equation on page 109. Each fringe moves a distance D in the time $1/\Delta\nu$, so they move at a speed $D\, \Delta\nu$.

Let us assume that the detection time for our eyes is about 10^{-3} sec. If the pattern flickered at the frequency $\Delta\nu$, this would mean that $\Delta\nu$ must be less than 10^3 Hz. The more patterned motion of the fringes that we will see might extend this somewhat, but the perception of motion is very like the perception of a flickering pattern. (Remember that enough detail must register to tell which direction things move.) So our simple criterion that $T_c > T_d$ is applicable, and the value of 10^3 Hz is a reasonable one. This, then, means that

$$v = \Delta\nu(c/\nu) = \lambda\, \Delta\nu \simeq 10^3\ \text{sec}^{-1} \cdot \frac{1}{2}\mu\text{m} = \frac{1}{2}\ \text{mm/sec}.$$

9.1 With four slits we will have phasors in the following configurations:

$\gamma = 2\pi m \quad (\sin\theta = m\lambda/d)$ → $E_T = 4E_i,\ I_T = 16\,I_i$

$\gamma = \pi/2,\ 3\pi/2 \quad (\sin\theta = \lambda/4d,\ 3\lambda/4d)$ → $E_T = 0,\ I_T = 0$

$\gamma = \pi \quad (\sin\theta = \lambda/2d$ → $E_T = 0,\ I_T = 0$

$\gamma = 2\pi/3,\ 4\pi/3 \quad (\sin\theta = \lambda/3d,\ 2\lambda/3d)$ → $E_T = E_i,\ I_T = I_i$

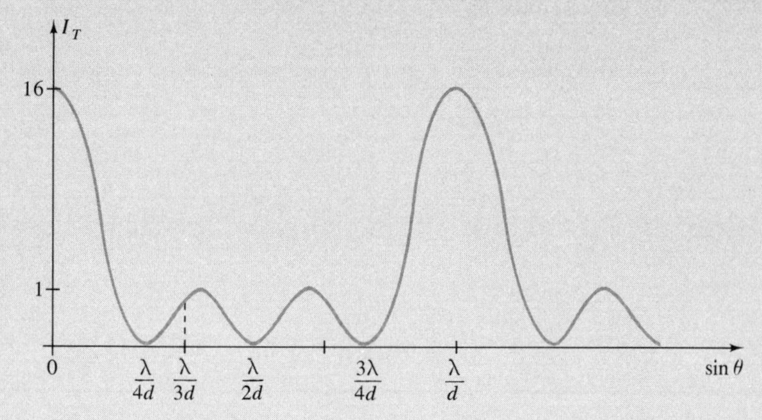

9.3 (a) We can use phasors to see that light from A and B must be $2\pi/3$ out of phase: $\gamma = 2\pi/3$.
To check this we can write:

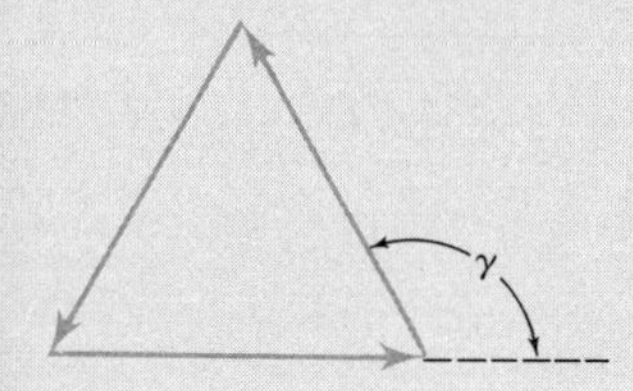

$$E_T = E_0 \sin\left[\frac{2\pi L_A}{\lambda} - 2\pi\nu t\right] + E_0 \sin\left[\frac{2\pi L_B}{\lambda} - 2\pi\nu t\right] + E_0 \sin\left[\frac{2\pi L_C}{\lambda} - 2\pi\nu t\right].$$

Take this at $t=0$ and $L_A = L_B + \lambda/3 = L_C + 2\lambda/3$:

$$E_T = E_0\left\{\sin\left(\frac{2\pi L_C}{\lambda} + \frac{4\pi}{3}\right) + \sin\left(\frac{2\pi L_C}{\lambda} + \frac{2\pi}{3}\right) + \sin\left(\frac{2\pi L_C}{\lambda}\right)\right\} = 0.$$

(b) Here $\gamma = \pi$; thus

E_C, E_B, E_A

$$E_T = E_0 = \frac{1}{3}(E_T)_{max}.$$

At the point C, we have maximum intensity:

$$E_T = (E_T)_{max} = 3E_0,$$

$$I_C = 9I_0.$$

where I_0 is the intensity that slit A alone would contribute there. Here,

$$E_T = E_0, \qquad I_P = I_0 = \frac{1}{9} I_{max}.$$

Therefore

$$\frac{I_P}{I_{max}} = \frac{1}{9}.$$

(c) At *any* principal maximum, $I_P/I_{max} = 1$.

(d) If there were no interference effects, the *intensities* would add and the screen would be *uniformly* illuminated at an intensity I_{av}. Each slit would have contributed $\frac{1}{3}I_{av}$, which is what we have been calling I_0. So,

$I_{av} = 3I_0$ incoherent, intensities add.

$I_P = 9I_0$ coherent, instantaneous amplitudes add.

The energy that has gone into making point P brighter than the average has come from the places where the pattern is darker than the average.

9.7 (a)

$$\sin\theta = \frac{m\lambda}{d} \qquad \text{position of the } m\text{th principal max.}$$

$$\sin\theta_3 = 3\cdot\frac{6400\text{ Å}}{d} \qquad d = \frac{1}{4000}\text{ cm}$$

$$\sin\theta_3 = 3\cdot\frac{0.64\times10^{-6}\text{ m}}{(1/4000)(\text{m}/100)} = 0.768$$

$$\theta_3 = \sin^{-1}(0.768) \cong 50° \approx 1 \text{ rad}.$$

(b) We know that $\Delta(\sin\theta) = m(\Delta\lambda/d)$, where $\Delta\lambda = \lambda_2 - \lambda_1$. $(m\,\Delta\lambda/d)$ is small; so we can call this

$$\Delta(\sin\theta) = \sin(\Delta\theta) = \tan(\Delta\theta)$$

Since the lens is close to the grating, the distance from grating to screen is approximately f_2. So

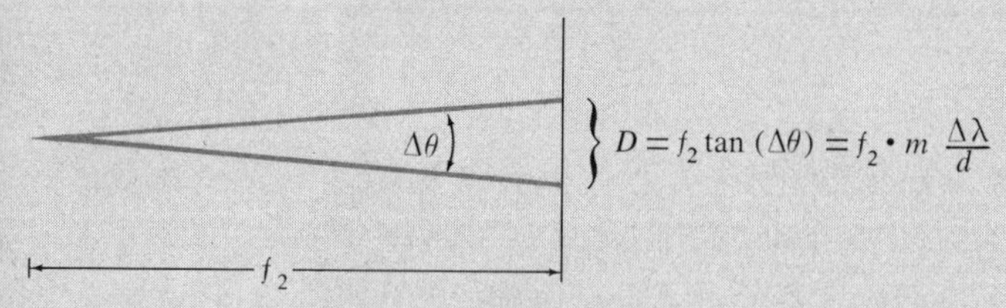

$$D = 3 \cdot f_2 \cdot \frac{\lambda_2 - \lambda_1}{d} = 3 \cdot 1\text{ m}\,\frac{0.5 \times 10^{-10}\text{ m}}{(1/4 \times 10^5)\text{ m}} = 0.06\text{ mm}.$$

(c) To find whether the lines are resolved in third order, we ask whether their separation is greater than their half-widths. We know that the answer will be "yes" if $\Delta\lambda > \lambda/Nm$. Here $N = 2000$, the number of lines illuminated in the grating.

$$\frac{\lambda}{Nm} = \frac{6400\text{ Å}}{3 \cdot 2000} = 1.06\text{ Å}.$$

$\Delta\lambda$ is less than this ($\Delta\lambda = 1/2$ Å), so the lines are not resolved. If we illuminated the whole grating, and went to fourth order, they would be resolved.

9.9 This is just a three-slit problem. The principal maxima come at $m(\lambda/d) = \sin\theta = m \cdot \frac{3}{2}$, so only the $m = 0$ principal maximum occurs. The first zero comes at

$$\sin\theta = \frac{\lambda}{3d} = \frac{1}{2} \qquad \left(\gamma = \frac{2\pi}{3}\right).$$

So the beam has an angular half-width of 30 degrees. We can find the intensity along xx from the phasor drawing:

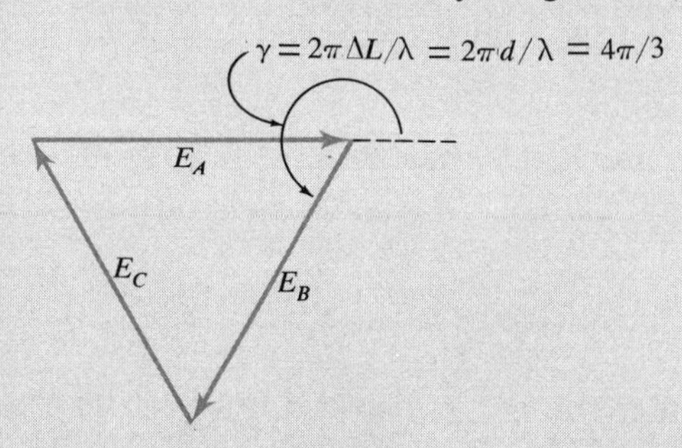

So $E_T = 0$, and the intensity along this line is zero. Our interference pattern is shown in the drawing, where we have plotted $I(\theta)$ on a *polar* plot: the larger *radial* values correspond to more intensity.

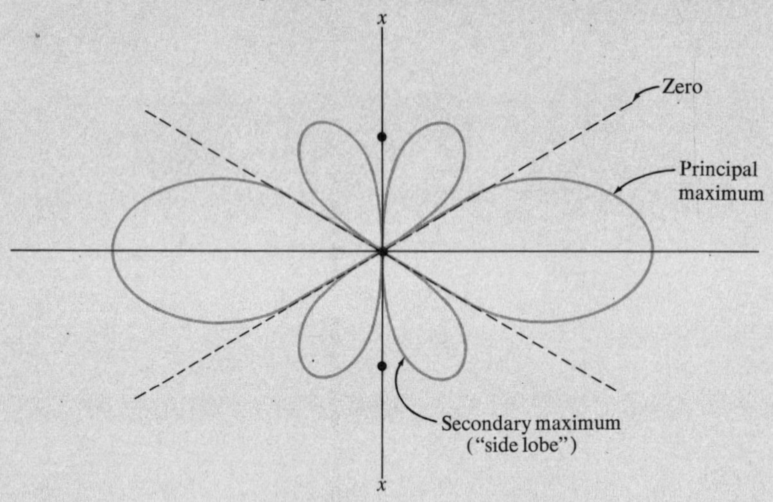

9.10 (a) This is a two-slit problem with slit spacing the same as in the original. The maxima also come at the same place.

(b) This is a two-slit problem with the same width of "grating" so that the maxima have the same width as in the original.

(c) Refer to Problem 9.1 for general features. Again, maxima coincide with those of the original.

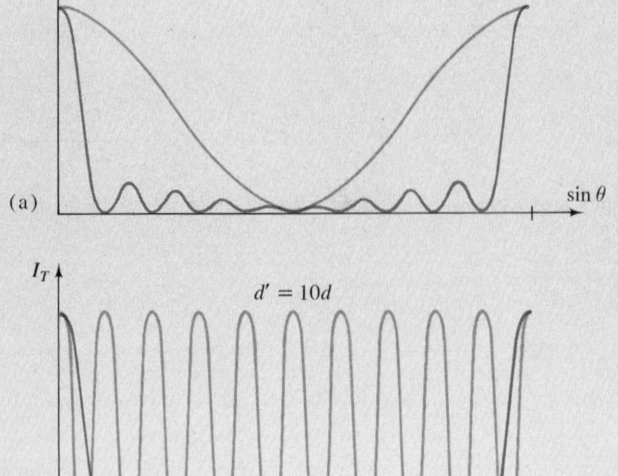

In the orginal, p_{max} *occurs at* $\sin\theta = m\lambda/d$, *and the zeros at* $\sin\theta = m'\lambda/10d$ $(m' \neq m)$. *In (a)* p_{max} *is again at* $\sin\theta = m\lambda/d$, *but the zeros are at* $\sin\theta = m'\lambda/2d$ $(m' \neq m)$. *In (b)* p_{max} *occurs at* $\sin\theta = m\lambda/d' = m\lambda/10d$, *while the zeros occur at* $\sin\theta = m'\lambda/2d' = m'\lambda/20d$ $(m' \neq m)$.

10.1 From the figure here we see that the dust and its image form two coherent sources, separated by a distance $2t$. First neglect refraction at the surface. At an angle θ from the normal, the two light paths differ in length by $2nt \cos\theta$, where n is the index of the glass. So the bright fringes occur at $2nt \cos\theta = m\lambda$. Since only θ is specified, all the rays with this angle form a cone of revolution about the normal, and the fringes appear circular. Close to $\theta = 0$ the fringe separation is large, but as θ increases, the fringes get closer together. Thus fringes are actually observable only close to the normal.

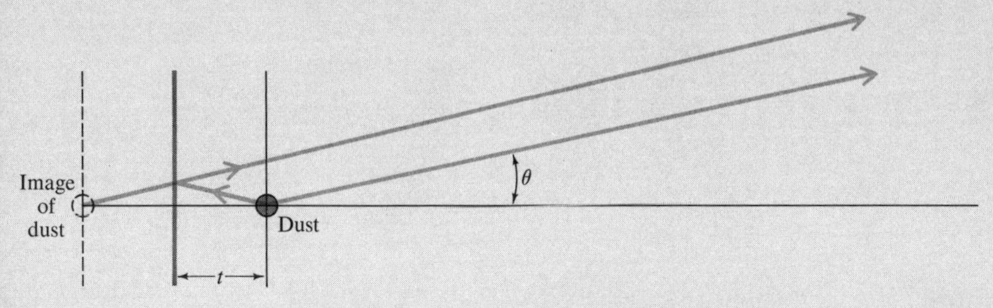

If we include refraction, the condition becomes $m\lambda = 2nt\sqrt{n^2 - \sin^2\theta}$.

10.4 Here bright fringes occur when the number of wavelengths in $2d_1$, minus the number in $2d_2$, is an integer.

$$\text{number of wavelengths in } 2d_2 = \frac{2d_2}{\lambda_{vac}},$$

$$\text{number of wavelengths in } 2d_1 = \frac{2(d_1 - t)}{\lambda_{vac}} + \frac{2t}{\lambda_{film}}.$$

We find a bright fringe when

$$\frac{2(d_1 - t - d_2)}{\lambda_{vac}} + \frac{2t}{\lambda_{vac}} \cdot n = m.$$

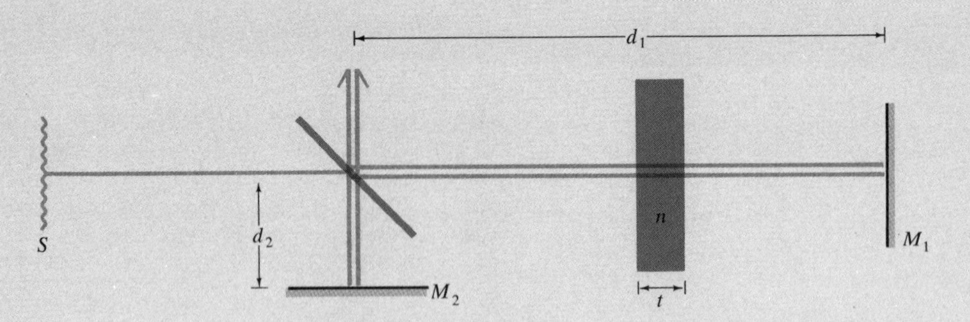

Before the film was present,

$$\frac{(2d_1 - d_2)}{\lambda_{vac}} = m - 7.$$

Subtracting,

$$7 = \frac{2t}{\lambda_{vac}}(n-1). \text{ So } t = 5.150\ \mu\text{m}.$$

A simple way to look at this is: taking out $2t/\lambda_{vac}$ wavelengths and putting in $2tn/\lambda_{vac}$ wavelengths resulted in seven more wavelengths. So $t(n-1) = 3.5\ \lambda_{vac}$.

10.5 If the curve $I(\lambda)$ is about like the one shown here, we can make an estimate by replacing it with two separate wavelengths at λ_0 and $\lambda_0 + \delta\lambda$. Then the fringes blur when a bright fringe at λ_0 coincides with a dark one at $\lambda_0 + \delta\lambda$:

$$m\lambda_0 = \left(m + \frac{1}{2}\right)(\lambda_0 + \delta\lambda) \qquad \text{or} \quad m \simeq \frac{\lambda_0}{2\delta\lambda}.$$

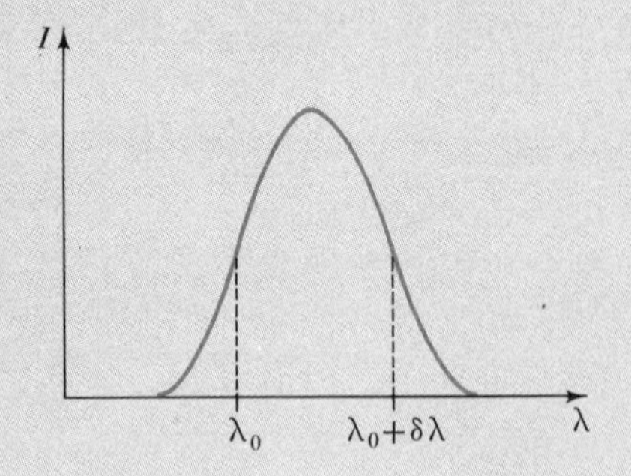

The departure from equal path lengths for this mth fringe is then $2d = m\lambda$.

Since path 1 can be either shorter or longer than path 2 by the distance d, we find for the range:

$$\Delta L_1 = \frac{\lambda_0^2}{2\delta\lambda}.$$

Wavelengths falling between λ_0 and $\lambda_0 + \delta\lambda$ will not be as much out of step; so the approximation primarily neglects those wavelengths in the "tails" of the spectral line.

The other way of looking at the problem is to say that the blurring of the fringes is due to the fact that the wave train is shorter than $2\,\Delta L_1$ and therefore, when the two parts are recombined at the beam splitter, one has an arbitrarily different phase from the other. This means that the coherence length is $L_c = 2\,\Delta L_1$, and the coherence time is

$$T_c = \frac{2\,\Delta L_1}{c} = \frac{\lambda_0^2}{c\delta\lambda} = \frac{\lambda_0}{\nu_0\,\delta\lambda} = \frac{1}{\delta\nu}.$$

10.6 First let us consider the intensities of the various beams. The upper beam, when it reaches A, has been reflected once and transmitted once, and so has an intensity $I = (0.9)(0.1)I_0$. The same is true for the lower beam, as observed at A. Thus observer A sees a superposition of two beams of equal intensity, and any dark fringes will be completely dark. Observer B, however, sees two beams of different intensity. The upper one has been transmitted twice, and so has intensity $I_u = (0.9)(0.9)I_0$, while the lower has intensity $I_e = (0.1)(0.1)I_0$ due to its double reflections. Since amplitudes will be proportional to the square root of intensities, the superposition seen by B will be of two waves of relative amplitude 1/9. The dark fringes there are never totally dark.

(a) Since mirror (2) is misaligned by δ, we have a situation like that of Problem 9.3. Observer A sees fringes of separation $D = (\lambda/2)/\sin(\alpha/2)$, as shown in the figure. Here the point $x = 0$ is arbitrary.

(b) Observer B sees a similar pattern, with less contrast.

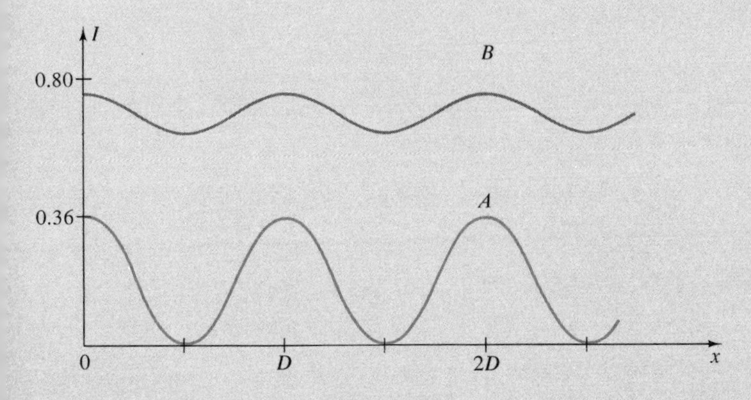

(c) Decompose the randomly polarized incident light into light polarized along horizontal and vertical axes. For the vertical component, the lower beam is retarded by π relative to the upper so that the fringes shift left (say) by $D/2$. What was a bright fringe becomes a dark one. Similarly, the fringes from the horizontal component shift right by $D/2$, so as a result, the whole pattern appears to reverse bright for dark. This would make a lot of difference if $\delta = 0$. Then a completely bright field would go completely dark.

10.8 (a) At the point of contact there is no film, so the reflection is *dark* (all light goes through). *Near* the point of contact, the film has negligible thickness, one reflection is internal ($\Delta\phi = 0$) and the other is external ($\Delta\phi = \pi$), so the reflection is *dark* there also, until $d = \lambda/4$.

$2d_{max} = 20\lambda$, so there are 21 dark fringes and 20 bright ones.

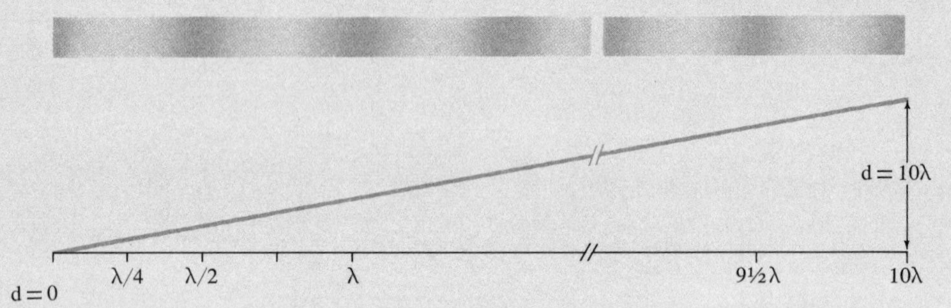

(b) $n = 1.8$ falls between the indexes of the two pieces of glass, so both reflections are now external. Thus we get

$$2d_{max} = 36\lambda_{liq} . \qquad \left(\lambda_{liq} = \frac{\lambda_{air}}{1.8}\right).$$

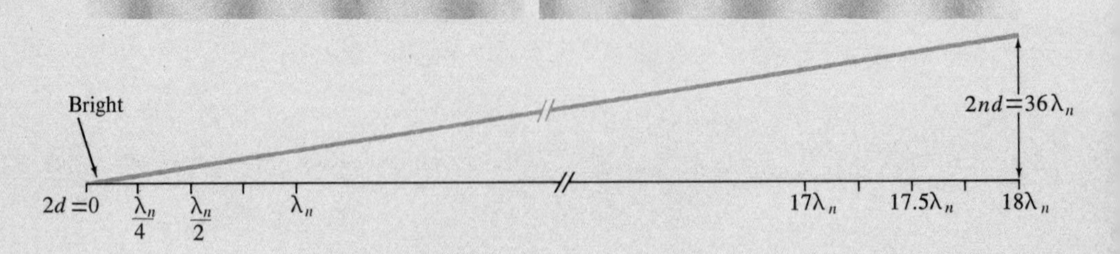

So we see 36 dark fringes and 37 bright ones.

10.11 Both reflections are "external", so destructive interference occurs when

$$2d = \left(m + \frac{1}{2}\right) \lambda_{\text{oil}} = \left(m + \frac{1}{2}\right) \frac{\lambda_{\text{vac}}}{1.3}.$$

Since d is always the same, only λ and m change.

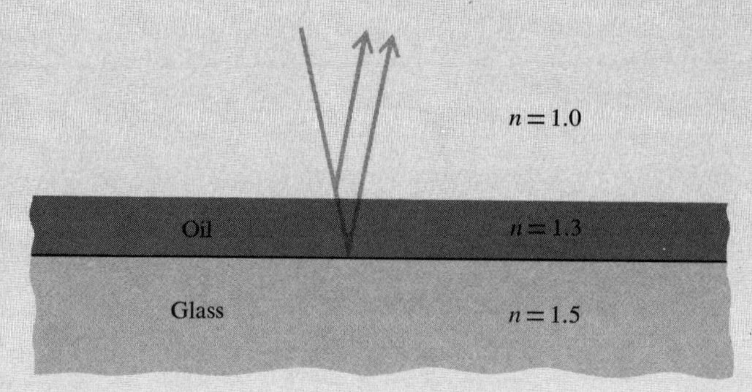

$$\left(m_1 + \frac{1}{2}\right) \frac{7000 \text{ Å}}{1.3} = \left(m_1 + 1 + \frac{1}{2}\right) \frac{5000 \text{ Å}}{1.3};$$

so

$$\frac{m_1 + \frac{3}{2}}{m_1 + \frac{1}{2}} = \frac{7}{5}, \qquad \frac{2m_1 + 3}{2m_1 + 1} = \frac{7}{5}, \qquad m_1 = 2$$

Then, going back to the condition for a dark fringe,

$$2d = \left(m_1 + \frac{1}{2}\right) \frac{7000 \text{ Å}}{1.3}, \qquad d = 6740 \text{ Å}.$$

11.2 All groups of rays that leave the slit as parallel "bundles" are brought to

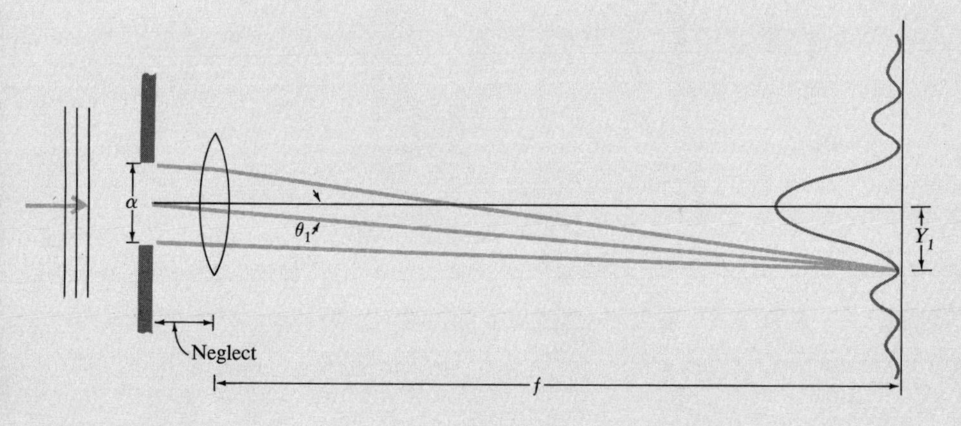

a focus by the lens. So we may follow the (undeviated) central ray of each bundle and use its θ. For instance, the first minimum occurs at $\sin\theta_1 = \lambda/a$:

$$Y_1 = f\sin\theta_1 = 900\text{ mm}\cdot\sin\theta_1$$
$$= 900\text{ mm}\left(\frac{\lambda}{a}\right)$$
$$= 1.0\text{ mm}.$$

11.4 Assume $\lambda = 6000$ Å; diameter of pupil, $a = 4$ mm; separation of lights, $S = 1.22$ m. Note that the sources are incoherent, so we find two single-aperture diffraction patterns, overlapping. We let them overlap until the principal maximum of one coincides with the first minimum of the other. This occurs when the lights subtend an angle α such that $\sin\alpha = 1.22\,\lambda/a$. We see that $\frac{1}{2}S/D = \tan\frac{1}{2}\alpha \simeq \frac{1}{2}\alpha \simeq \frac{1}{2}\sin\alpha$. So $S/D = 1.22\lambda/a$, or $D = 1.22\text{ m}\cdot 4\text{ mm}/(1.22\cdot 6\times 10^{-4}\text{ mm}) = 6.6$ km.

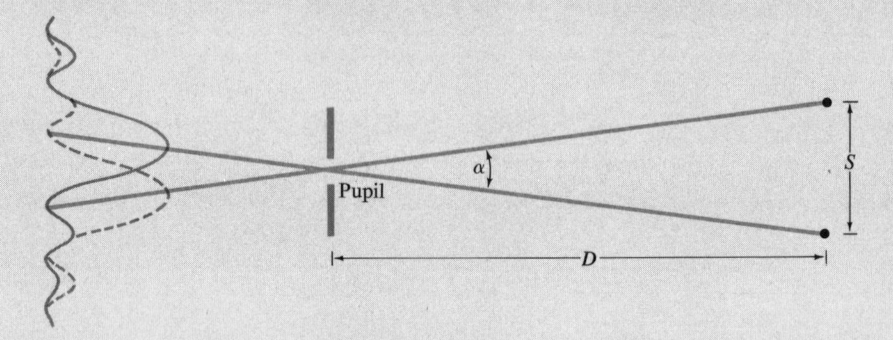

This is an overestimate, since defects in the eye, turbulence in the air, and finite size of the lights are all detrimental to the "seeing", in the astronomers' sense of the word.

11.6 Highest accuracy is obtained by measuring between the most widely separated minima possible. In the picture this might mean from $m = -3$ to $m = +3$.

$$\sin\theta_{+3} = \frac{+3\lambda}{a}, \qquad \sin\theta_{-3} = \frac{-3\lambda}{a}. \qquad \text{Difference} = 6\times\lambda/a.$$

$$\tan\theta_3 = \frac{Y_3}{D} \simeq \sin\theta_3.$$

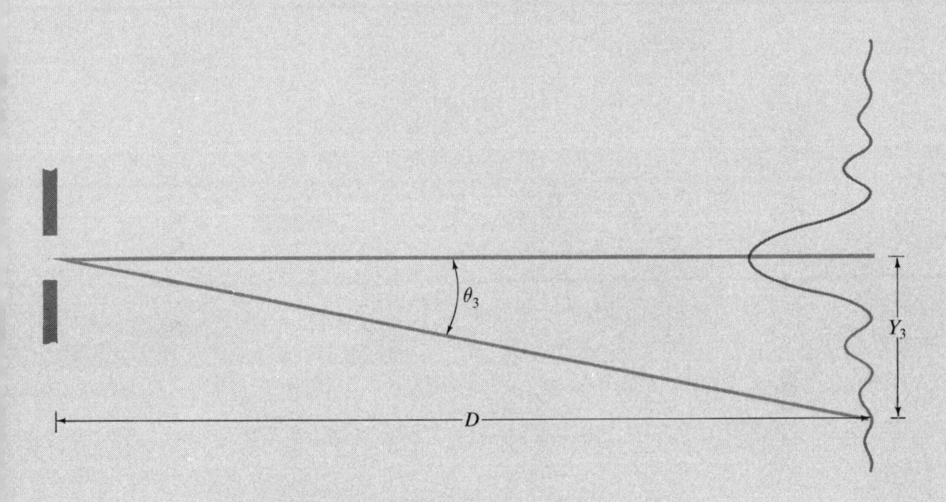

We measure $2Y_3 = 20$ mm ± 0.5 mm:

$$\frac{2Y_3}{2m} = \frac{6(6328\ \text{Å})}{a}. \qquad a = 3.6\ \text{mm} \pm 0.09\ \text{mm}.$$

11.8 Zeros occur at $\sin\theta = m\lambda/a$. At the angle θ_I, $\sin\theta_I = \lambda_a/a = 2\lambda_b/a$, so $\lambda_a = 2\lambda_b$. Clearly, the coincidence occurs again at θ_{II}, where $\sin\theta_{II} = 2\sin\theta_I$, and at all further multiples.

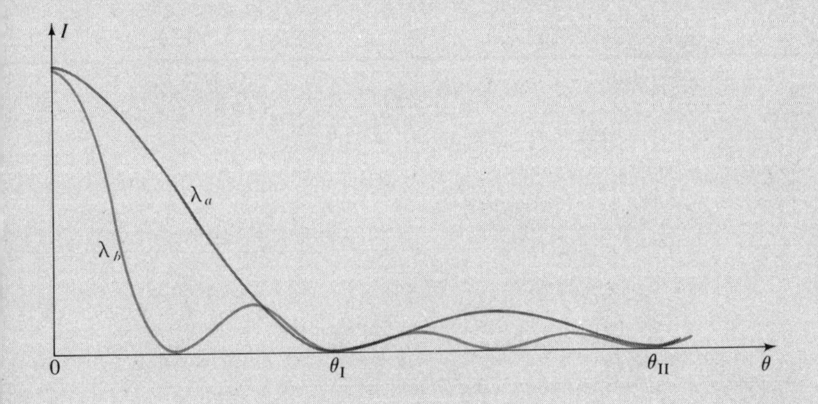

11.10 The half-width is equal to the distance from the center to the first minimum: $\frac{1}{2}$ width $= \lambda/a = \sin\theta_1$.

(a) $a = \lambda$, $\quad \sin\theta_1 = 1$.

$$\theta_1 = \frac{\pi}{2}$$

$$\frac{1}{2}\text{ width} = \frac{\pi}{2}\text{ rad} = 90°.$$

(b) $a = 5\lambda$, $\sin\theta_1 = \dfrac{1}{5}$.

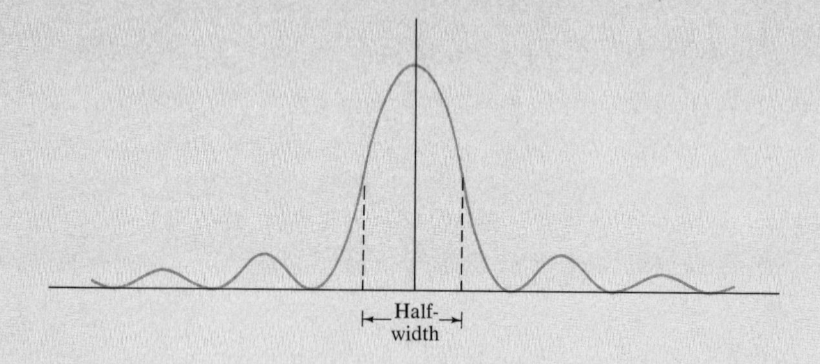

$\theta_1 = 11°30' = 0.20$ rad.

(c) $a = 10\lambda$, $\sin\theta_1 = 1/10$.

$\theta_1 = \frac{1}{10}$ rad $= 5°45'$. (A "tight" beam.)

11.12 The simplest way to describe this is to change to a new wavelength, $\lambda' = \lambda/n$. Since we measure everything in terms of λ, we will now find that

$$I(\theta) = I_0 \frac{\sin^2(\pi\lambda/na \sin\theta)}{(\pi\lambda/na \sin\theta)^2},$$

zeros at $\sin\theta = m(\lambda/na)$. We could have scaled the slit, changing a to na, but the distance to the screen does not change so that this might be misleading. To be entirely safe, we could derive the intensity, as in the $n = 1$ case, and use $\lambda = \lambda'/n$. This would, of course, lead to the answer given above.

12.2 We could phrase this question in this way: What object has a diffraction pattern like the screened-off one? The unblocked pattern is that which we would have obtained from crossed gratings of half the spacing, so it is this image that we see.

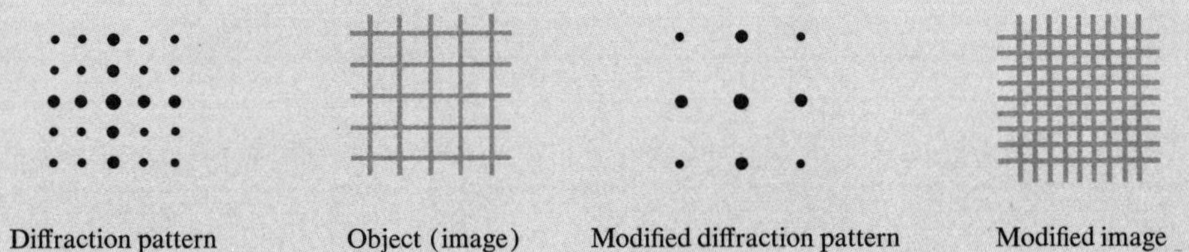

12.4 If the two beams differ by the angle α, and both are plane waves, as in Problem 11.1, the maxima are a distance d apart, where $d = (\lambda/2)\cot\alpha/2$. Since the value of d can be no smaller than $d = 1\ \mu\text{m}$, the value α_{max} is

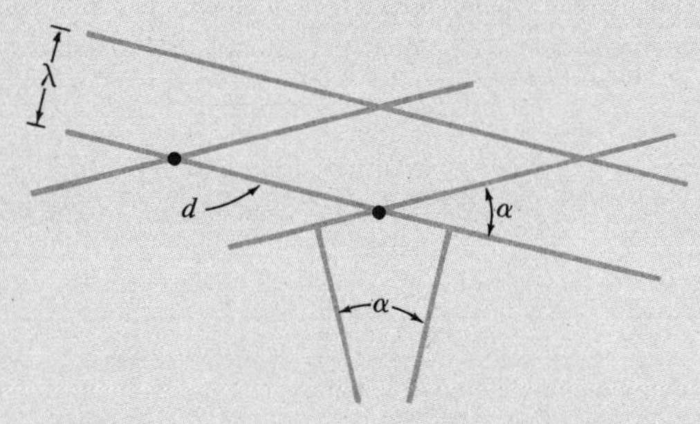

$$\cot\frac{\alpha_{max}}{2} = \frac{1\ \mu\text{m}\cdot 2}{0.6328\ \mu\text{m}} = 3.16 \qquad (\alpha_{max} \simeq 0.6328\ \text{rad} \simeq 36°).$$

In the case of a diffusely scattering object, then, we have the geometry shown, from which: $L = 2D\tan\alpha_{max} = 4.218$ cm.

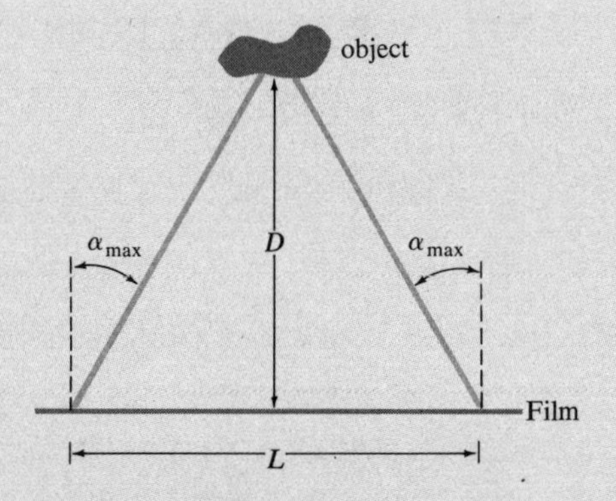

Index

G

H

I

K

L